Springer Series in Fashion Business

Series editor

Tsan-Ming Choi, The Hong Kong Polytechnic University, Hung Hom, Hong Kong

More information about this series at http://www.springer.com/series/15202

Tsan-Ming Choi
Editor

Analytical Modeling Research in Fashion Business

Springer

Editor
Tsan-Ming Choi
Institute of Textiles and Clothing
The Hong Kong Polytechnic University
Hung Hom
Hong Kong

ISSN 2366-8776 ISSN 2366-8784 (electronic)
Springer Series in Fashion Business
ISBN 978-981-10-1012-5 ISBN 978-981-10-1014-9 (eBook)
DOI 10.1007/978-981-10-1014-9

Library of Congress Control Number: 2016938408

Printed on acid-free paper

This Springer imprint is published by Springer Nature
The registered company is Springer Science+Business Media Singapore Pte Ltd.

Preface

The fashion industry is one of the most important industries in the world. Traditional studies and research on fashion business are mostly empirical in nature, e.g., case studies and empirical analyses. However, with the development of the "fashion business" field of studies and the need for deeper theoretical foundations, an expansion of research methodology is naturally needed. In fact, nowadays, there are more and more related publications which employ an analytical approach in conducting theoretical and applied research in fashion business.

However, in the literature, there is no comprehensive reference source that provides the state-of-the-art findings on both theoretical and applied analytical research related to fashion business. It is thus significant to put together some interesting works and the respective insights into an edited volume. To be specific, this handbook consists of three important parts which include: (i) reviews and discussions, (ii) theoretical economics models, (iii) engineering models, applications, and cases. The specific topics covered include the following:

- Introduction to analytical modeling research in fashion business.
- Analytical modeling research methodologies in fashion business operations management.
- Consumer returns in fashion retailing.
- The role of quick response strategies in accelerating fashion sales.
- Using mixed channels for fashion apparel retailing.
- Vendor-managed inventory partnerships with markdown money supply contracts.
- Inventory management in fashion retailing with a random supply.
- Optimal fashion sourcing, quotation and in-house production decisions.
- Distribution, transshipment, and sustainable logistics management for fashion business.

- Order-picking systems by third-party logistics service providers for fashion firms.
- Inventory control with service level and lead time considerations.
- An analytic hierarchy process (AHP) scheme for enhancing fashion sales forecasting.

As a researcher who has conducted analytical modeling research in fashion business myself over the past two decades, I am very pleased to see that this handbook contains new discussions on research methodologies, and important research findings and insights, which contribute significantly to the literature and help advance both applied and academic research in fashion business.

Before closing, I would like to take this opportunity to thank William Achauer (Bill) of Springer for his constructive advice along the course of carrying out this important book project, and his kind support for the establishment of Springer's Book Series in Fashion Business. I am indebted to all the authors who have contributed their interesting research to this handbook, and the anonymous reviewers who have provided timely and helpful reviews to the papers. I am also thankful to my research team members Hau-Ling Chan, Shu Guo, and Shuyun Ren for their assistance. Last but not least, I would like to dedicate this book to my parents and my family.

Hung Hom, Hong Kong Tsan-Ming Choi
December 2015

Contents

Contributors

Osman Alp Haskayne School of Business, University of Calgary, Calgary, AB, Canada

Feyzan Arikan Industrial Engineering Department, Faculty of Engineering, Gazi University, Maltepe, Ankara, Turkey

Aysenur Can Industrial Engineering Department, Faculty of Engineering, Gazi University, Maltepe, Ankara, Turkey

Chen Chen Rowe School of Business, Dalhousie University, Halifax, NS, Canada

Jing Chen Rowe School of Business, Dalhousie University, Halifax, NS, Canada

Tsan-Ming Choi Business Division, Institute of Textiles and Clothing, The Hong Kong Polytechnic University, Kowloon, Hong Kong

Mack A. Diggins Rowe School of Business, Dalhousie University, Halifax, NS, Canada

Ciwei Dong School of Business Administration, Zhongnan University of Economics and Law, Wuhan, China

Xu Guan School of Economic and Management, Wuhan University, Wuhan, Hubei, China

Zhi-Hua Hu Logistics Research Center, Shanghai Maritime University, Shanghai, China

Yu-chen Kang School of Management and Economics, Beijing Institute of Technology, Beijing, China; Center for Energy and Environmental Policy Research, Beijing Institute of Technology, Beijing, China

Alejandro Lago IESE Business School, University of Navarra, Barcelona, Spain

Guo Li School of Management and Economics, Beijing Institute of Technology, Beijing, China; Center for Energy and Environmental Policy Research, Beijing Institute of Technology, Beijing, China

Qingying Li Glorious Sun School of Business and Management, Donghua University, Shanghai, China

Yu Liu College of Computer Science and Technology, Wuhan University of Science and Technology, Wuhan, Hubei, China

Victor Martínez-de-Albéniz IESE Business School, University of Navarra, Barcelona, Spain

Philip Moscoso IESE Business School, University of Navarra, Barcelona, Spain

Samar K. Mukhopadhyay Graduate School of Business, Sungkyunkwan University, Jongno-Gu, Seoul, South Korea

Rongrong Qian Glorious Sun School of Business and Management, Donghua University, Shanghai, China

Vincent Quan Jay and Patty School of Business and Technology, Fashion Institute of Technology, State University of New York, New York, USA

Bin Shen Glorious Sun School of Business and Management, Donghua University, Shanghai, China

Tarkan Tan School of Industrial Engineering, Eindhoven University of Technology, Eindhoven, The Netherlands

Andreu Vall IESE Business School, University of Navarra, Barcelona, Spain

Xiaohang Yue Sheldon B. Lubar School of Business, University of Wisconsin-Milwaukee, Milwaukee, WI, USA

Chunnan Zhang Department of Accounting, Temple University, Philadelphia, PA, USA

Ying Zhang Department of Industrial and Systems Engineering, The State University of New York at Buffalo, Buffalo, NY, USA

Xiaowei Zhu College of Business and Public Affairs, West Chester University of Pennsylvania, West Chester, PA, USA

Ruixin Zhuang China Foreign Exchange Trade System, Shanghai, China

Part I
Reviews and Discussions

Chapter 1
Analytical Modeling Research in Fashion Business: An Introduction

Tsan-Ming Choi

Abstract The fashion industry is one of the most important industries in the world. Traditional studies and research on the fashion business are mostly empirical in nature, e.g., via case studies and empirical analysis. However, there is a need for establishing solid theoretical foundations to support the development of fashion business into a strong academic field. In this introductory chapter, we first briefly outline the topic on analytical modeling research in fashion business. We then concisely introduce the papers featured in this book in three parts, namely reviews and discussions; theoretical economic models; and engineering models, applications, and cases. Future research directions are discussed.

Keywords Analytical modeling research · Fashion business · Introduction · Review

1.1 Introduction

The fashion industry, which includes fashion apparel, fashion beauty and cosmetics, footwear, fashion accessories, watches, jewelries, etc., is a huge industry. Traditional research in fashion business focuses on exploring the problems via case studies or by applying other empirical approaches (such as survey, interviews, and some statistical data analysis). However, when time passes, there is a need in terms of methodology development for the field of fashion business to advance. The well-established analytical modeling research methodology, with its roots in prominent scientific fields of economics and engineering, becomes the right candidate to fulfill this need.

T.-M. Choi (✉)
Business Division, Institute of Textiles and Clothing, The Hong Kong Polytechnic University, Hung Hom, Kowloon, Hong Kong
e-mail: jason.choi@polyu.edu.hk

T.-M. Choi (ed.), *Analytical Modeling Research in Fashion Business*,
Springer Series in Fashion Business, DOI 10.1007/978-981-10-1014-9_1

In fact, analytical modeling research focuses on deriving scientifically sound findings from the development of mathematical models. In the scope of fashion business, conducting analytical modeling research can both enhance industrial practices and enrich the respective theories in the literature. In the following, we concisely review some important papers in fashion business. First, in the domain of fashion marketing and consumer behaviors, noticing that conspicuous consumption is a critical issue in markets such as luxury fashion, Zheng et al. (2012) analytically investigate the optimal advertising and pricing decisions for a luxury fashion brand with the considerations of social influences between different groups of consumers. The authors interestingly find that it is indeed optimal for the luxury fashion brand to allocate resources to focus on only one group of consumers. After that, Tereyagoglu and Veeraraghavan (2012) study the pricing and inventory decisions for a retailer when consumers exhibit conspicuous characteristics. The authors consider both the homogeneous and heterogeneous market scenarios and generate insights by comparing the equilibrium decisions between the cases when the consumers are strategic and the consumers are conspicuous. Other analytical modeling studies related to fashion marketing and consumer behaviors include the examination of the fast fashion strategies (with quick response and enhanced design components) in the presence of strategic consumers (Cachon and Swinney 2011), the dynamic e-pricing study for fashion retailers with the use of electronic price testing (Choi et al. 2012), the optimal service charging scheme for consumer returns under fashion mass customization programs (Choi 2013b), and the optimal advance selling strategy of fashion products in the presence of opportunistic consumers (Li et al. 2014b). In the scope of fashion supply chain operations management, quite a lot of studies published in the literature employ the analytical modeling approach. For instance, Eppen and Iyer (1997a) study the backup agreement in fashion supply chains. Eppen and Iyer (1997b) discuss the use of Bayesian information updating for improving fashion buying operations. Iyer and Bergen (1997) investigate the channel coordination issue in quick response apparel supply chains. Donohue (2000) explores the fashion supply chain coordination problem by using a special wholesale pricing and returns contract in the presence of forecast information updating. Petruzzi and Monahan (2003) examine the use of secondary outlet stores in fashion retail inventory management. Choi (2007) studies the pre-season inventory and pricing policies for fashion retailers. Other related fashion supply chain operations management studies include Chiu et al. (2011), Wu and Chiang (2011), Yang et al. (2011), Choi (2013a, c, d, e, f), Shen et al. (2013), Li et al. (2014a), Xiao et al. (2014) and Wang et al. (2015). Observe that in the above-reviewed analytical modeling research, the focal points are on developing managerial insights, and hence, the models employed are relatively simple and there are quite a number of assumptions imposed on the models and the corresponding analyses (e.g., many of them only consider the presence of a single product). In this paper (and also in this book), these studies are collectively classified as "economic analytical modeling research."

There is another kind of analytical modeling research which focuses on developing implementable tools (e.g., developing the tools as applicable decision supporting systems) and revealing insights from the respective applications. We classify them as the "engineering analytical modeling research." In the literature, we can find a number of fashion business-related papers on engineering analytical modeling research. For example, Yu et al. (2011) propose an intelligent collaborative scheme for enhancing the fashion design process in practice. By using real data, they examine the performance of their proposed method and argue that it is highly efficient in generating high-quality fashion designs within a reasonably short period of time. Caro and Gallien (2012) discuss the clearance pricing optimization problem in the fast fashion retailer Zara. They report the real implementation of their proposed model in Zara and demonstrate (via field experiments) its significance in lifting the clearance revenue. Choi et al. (2014) explore the fast fashion sales forecasting system in which there is only a limited amount of historical data available. Their system includes different sub-modules which can provide a good forecasting accuracy within the given short computational time constraint. Other fashion business-related engineering analytical modeling studies include Yeung et al. (2010), Caro and Gallien (2010), Bucci et al. (2014), Hu and Yu (2014), Hu et al. (2015) and Jia et al. (2015).

Motivated by the importance of analytical modeling research in fashion business, this research handbook is compiled. In the following, we briefly introduce each featured paper in this handbook in three categories: (1) Reviews and discussions; (2) Theoretical economic models; and (3) Engineering models, applications, and cases. After that, we discuss the future research in the related areas.

1.2 Reviews and Discussions

First of all, as there is no formal published guideline on the proper research methodology for analytical modeling research in fashion business operations management, in Chap. 2, Choi fills this gap by exploring and proposing two research methodologies. The author first concisely reviews the related literature on operations management analytical modeling research. He then proposes to classify the operations management analytical modeling research, in the context of fashion business, in two categories, namely the economic analytical modeling research and the engineering analytical modeling research. For each category, he proposes the respective steps for accomplishing rigorous research in fashion business operations management problems. For each step, he further presents some related important remarks specific to fashion business operations. He concludes by showing the comparisons between the two proposed categories of analytical modeling research methods, and giving comments on the proposed classification scheme.

Customer returns are becoming a common phenomenon in the retail industry. Due to the unique characteristics of fashion retailing, including short-product life cycle, high returns rates, fraudulent returns, and low salvage value, customer returns become one very serious issue faced by fashion retailers. In fact, customer returns are growing at a significant rate every year, costing the retailing industry billions of dollars. It is interesting to observe that the customer return rate differs among different types of industries while it is very significant in the fashion retailing industry. Furthermore, fashion products purchased online are generally done conveniently, which also leads to high returns since customers are unable to have the "feel and touch" service when they make purchasing decisions. Overall speaking, even though fashion retailers can achieve competitive advantages by offering free shipping and returns, numerous problems arise and the associated costs accumulate. Therefore, the fashion retail industry is finding customer returns management a big challenge. Motivated by the above-reviewed industrial needs, in Chap. 3, Diggins, Chen and Chen review and explore important issues related to customer returns management in fashion retailing. In fact, Diggins et al. argue that academic researchers have paid attention to customer returns and already identified customer returns as an important aspect of the reverse logistics process. There have been, however, relatively few studies exploring the unique impact of customer returns on the fashion retailing industry. Customer return policies are very important in fashion retailing since it influences consumer behaviors to make the initial purchase by reducing any uncertainty or risk associated with an unfamiliar fashion product. The biggest issue arises when customers take advantage of a retailer's return policy and then commit return fraud when returning items with no intention of ever keeping the products. Products being returned in the fashion industry impose great risk to the fashion seller because of the short-product life cycles and unpredictable market demand. Another challenge that is unique to the fashion industry is the concept of applying customer return policies to products that are produced using mass customization techniques. Since these products are customized to an individual's preferences, there is great difficulty in reselling them once they are returned, therefore reducing the returned products' salvage value. Furthermore, having liberal return policies accompanied with fashion products has become evident in fashion e-commerce for competitive and value creation purposes. In Chap. 3, the authors review the literature considering the impacts of customer returns policies, methods of reducing customer returns, and management decisions related to customer returns, as these topics are all related to or can be applied to the fashion retailing industry. With customer returns receiving more attention from both the practitioners and the academics, the authors have uncovered many open research gaps and opportunities for future research.

1.3 Theoretical Economic Models

This book features four papers which are classified under the category of "theoretical economic models." The first one, which appears in Chap. 4, is an interesting paper by Lago, Martinez-de-Albeniz, Moscoso, and Vall. Quick response, a supply chain management strategy initiated in the American fashion apparel industry back to the 1980s, is well-advocated to deal with demand uncertainty. The key spirit of quick response is to reduce lead time so that many supply chain processes can be postponed to a time point when demand uncertainty is significantly reduced. In Chap. 4, Lago et al. study the impact of lead time and sourcing origin on fashion product's success. The authors construct an analytical sales diffusion model in which the product success is characterized by the speed of sales. By employing the real data provided by a European-based fast fashion retailer, they evaluate how the speed of sales is affected by the product design time and the time to market of every product item. They reveal that design postponement, which refers to delaying the time of design, can bring significant benefit to the fast fashion retailer. They also find that the influence brought by the time-to-market factor is mixed with other factors. For example, when the time to market is longer, the speed of sales becomes lower while the degree of improved learning from design postponement is higher. The authors discuss several important managerial implications of the findings regarding quick response practices and propose future research directions.

More and more companies are opening direct channels for selling fashion clothes to customers, which are in addition to their more traditional bricks-and-mortar offline retailing channel. Inspired by the above industrial observation on mixed-channel retailing, Zhu, Mukhopadhyay, and Yue present in Chap. 5 an analytical model to explore the optimal pricing decisions by the manufacturer and the retailer in the respective mixed-channel supply chain. The authors explore the amount of "value-adding" provided by the offline retailer and determine the manufacturer's optimal wholesale price to the retailer. Their model considers information asymmetry where the manufacturer has incomplete information about the offline retailer's cost of adding value. They employ a game-theoretic formulation for this mixed-channel retailing problem and consider three different scenarios, including the "channel integration (I)" scenario, the "full information (F)" scenario, and the "asymmetric information (A)" scenario. The authors obtain closed-form equilibrium contracts for these three cases and compare the results by using both analytical and numerical analyses. They develop a number of managerial guidelines for fashion companies. In particular, they argue that information asymmetry imposes inefficiency on the manufacturer and on the supply chain as a whole. The channel-integration scenario is always the first best case. The retail channel price under the channel-integration scenario gives the lowest retail selling price among all the three cases. The retail channel prices are increasing with the channel migration factor for all the three cases of (I), (A), and (F). The authors further propose that since customers prefer to visit a physical offline retail store to shop the high-end fashion and luxury brand products, the manufacturer should encourage the respective retail store to add more value-added services. Customers prefer to shop online for fast fashion clothes. The fast fashion brand should open a direct channel in addition to the retail store and let two channels compete with each other. Fashion companies can also consider differentiating products to reduce the degree of channel conflicts. The authors also identify several promising research topics for future studies.

A vendor-managed inventory (VMI) partnership is a program well-recognized as one of the most successful practices in the fashion industry. Undoubtedly, a proper setup VMI can enhance strategic cooperation and improve efficiency throughout the supply chain. Motivated by real industrial practices in the fashion industry, Shen Qian and Quan investigate in Chap. 6 the issue of how a VMI partnership with the markdown money policy (MMP) operates in the fashion supply chain. Under the observed MMP, a vendor allowance (VA) in the form of markdown money sponsor is issued from the supplier to the said retailer every quarter or six months per season. The authors analytically examine how an MMP affects supply chain performance. Furthermore, a numerical study is included to uncover the impacts of VMI within the supply chain in the presence of the MMP. Implications to fashion industrial practices are discussed.

Random supply is a challenging issue for fashion retailers. To address this issue, Li, Dong, and Zhuang consider a fashion retailing inventory problem in Chap. 7. To be specific, they investigate a situation in which the fashion retailer, which is modeled by the standard newsvendor model, needs to decide the optimal ordering quantity for a fashion item from an unreliable fashion supplier. Here, the term "unreliable" means the supply is random. To generate analytical tractable results, the authors formulate the retail selling price of the fashion product as a decreasing function of the supplier's random supply. The authors further examine two different cases, namely the procurement case (where the fashion retailer is just a retailer) and the in-house production case (where the fashion retailer both produces and sells the product itself). Analytical comparisons of the optimal ordering quantities are reported. Results from sensitivity analysis are shown and managerial insights are revealed.

1.4 Engineering Models, Applications, and Cases

In addition to the theoretical economic modeling studies, this book also includes five papers which focus on applications and cases. We call them the "engineering analytical modeling" research and introduce them as follows.

First, Tan and Alp investigate in Chap. 8 the sourcing decisions of a fashion product manufacturer. The authors examine three major issues, namely the supplier selection problem from the manufacturer perspective, the supplier's capacity and price quotation problem, and the manufacturer's outsourcing or in-house production decision. They model the problem with a stochastic demand and capacitated production facilities. They develop a stochastic dynamic programming model to solve the problem. They discuss pricing and channel coordination challenges. They also generate several important insights, which include: (i) They reveal that a more flexible sourcing system may yield a reduction of the total quantity ordered. (ii) The optimal sourcing solution is not necessarily robust. (iii) The total quantity ordered may decrease as variability of demand increases. (iv) It is important to integrate the supplier selection and the procurement decision together as the result yields significant impacts.

In Chap. 9, Hu examines the distribution management, transshipment control, and sustainable logistics management in the fashion industry. In the first problem, motivated by the fact that customers treasure the "try-on service", the author first discusses the try-on service and builds the related analytical model. By solving the optimization problem, he reveals the try-on service's effects on customer return ratios, customer loyalty, and logistics costs. In the second problem, the author explores the fashion transshipment control problem. He shows the complexity of the problem and discusses its applications. In the third problem, the author discusses the timely sustainable logistics management issues. Finally, the challenges and opportunities in logistics management for the fashion industry in the presence of advanced technologies such as the Internet of Things and big data are explored.

Warehouse operations are critically important in logistics management of a fashion retail supply chain. Although the literature contains a huge amount of scientific studies on the decision problems for the design and control of warehouses, there are only limited studies on warehouse operations in fashion companies. Among the warehouse operations, order picking is the most labor-intensive and costly activity for almost every warehouse. Hence, it is the highest priority activity to improve the company's productivity. In Chap. 10, Can and Arikan propose a process flow analysis for the design of order-picking systems by integrating the strategy selection problem into it with detailed implementation steps and utilizing these steps simultaneously. Implementation steps include a mathematical modeling-based analytical approach to determine the optimal order-picking strategy by a proper cost-centric evaluation scheme. To calculate the cost of different systems corresponding to each order-picking strategy, the authors propose a model which is revised from the literature. Furthermore, this study provides a good guide for warehouse managers to execute the implementation steps of the proposed design flow for real-life applications. The process flow can be adapted to any order-picking design and selection problem with slight modifications according to the requirements of the considered system.

The fast fashion trend, which encourages high circulation speed and operations efficiency, requires products to be designed and manufactured quickly and inexpensively. Actually, this philosophy of quick manufacturing at an affordable price has permeated in large retailers such as H&M, Gap, and Zara. Driven by the fast fashion practices, it is crystal clear that lead time and service level are both critical factors in fashion retail operations. In Chap. 11, Li, Kang, and Guan explore fashion supply chain management through an analytical modeling optimization approach. Specifically, the authors develop an inventory optimization model subject to the controlled lead time and service level constraint, and provide its optimal analytical solution. The model yields the optimal order quantity decisions of the buyer, the optimal production quantity and the optimal lead time of the vendor with the service level constraint set in the fashion supply chain. The model also allows the cost of compressing lead time to have a linear relation with the segmented lead time. The authors examine a minimax method to calculate the value of the optimal products which is delivered from the vendor to the buyer. They discuss the mutual influence between the variables and the robustness of the analytical results using numerical analyses.

In Chap. 12, Zhang et al. examine an analytical hierarchy process (AHP)-based aggregation–disaggregation scheme for improving sales forecasting in the fashion industry. The authors employ real empirical sales data from a fashion boutique for the analysis. They propose a novel AHP-mediated scheme which first aggregates the historical sales data and then disaggregates the forecast quantities. They incorporate their proposed AHP scheme into the moving average and the exponential smoothing time series forecasting techniques. Their real data-based experimental results show that their proposed AHP scheme can help improve sales forecasting for products with low sales volumes and limited historical data. They also discuss several important future research areas.

1.5 Concluding Remarks, Research Agenda and Acknowledgments

In this introductory chapter, we have concisely discussed the topic on analytical modeling research in fashion business. First, some recently published works employing the "economic analytical modeling approach", in fashion marketing, and fashion business operations management have been reviewed. After that, we have examined the literature on papers focusing on developing tools, real-world applicable solutions, and cases and we have classified them as the papers following the "engineering analytical modeling approach". Then, we have introduced the papers featured in this book in three sections, namely Sect. 1.2 on reviews and discussions; Sect. 1.3 on theoretical economic models; and Sect. 1.4 on engineering models, applications, and cases. As a summary, Table 1.1 shows the topic examined by each chapter in this book and the corresponding proposed future research directions. These proposed future research directions are all meaningful and provide constructive suggestions for specific further studies. We strongly believe that the findings of this book and proposed future research topics will help stimulate new analytical modeling research in fashion business.

Table 1.1 Topics and future research directions of the book chapter papers featured in this book

Chapters	Topics	Future research directions
2	Analytical modeling research methodologies for fashion business operations management	a. Refine and improve the proposed methodologies b. Explore other probable research methodologies such as the analytical modeling research methodology with behavioral experiments, and some other multi-methodological research approaches
3	Consumer returns in fashion retailing	a. Identify measures to reduce consumer returns b. Explore the consumer behaviors in returning fashion products c. Determine the methods to identify fraudulent returns in fashion d. Study how consumer satisfaction and loyalty affect product returns e. Investigate the interrelationship between the consumer returns policy and other marketing mix decisions
4	The role of quick response in accelerating fashion sales	a. Extend the analysis to other kinds of fast fashion operations b. Conduct more analytical studies which explicitly model demand dynamics related to inventory availability
5	Using mixed channels for apparel retailing	a. Extend the study by incorporating advertising strategy into the model b. Study the omni-channel integrated retailing model
6	VMI partnerships with markdown money policies	a. Extend the study to include sales effort decisions b. Consider the use of more sophisticated contracts under VMI
7	Inventory management in fashion retailing with random supply	a. Extend the model to include a stochastic price-dependent demand process b. Consider the risk-sensitive optimization objectives in the inventory control models c. Explore the case when the random supply and the random demand are correlated
8	Optimal fashion sourcing, quotation and production mode decisions	a. Examine the multi-period decision-making case b. Explore the case with supply disruptions c. Incorporate multiple criteria into the optimization model
9	Distribution, transshipment and sustainable logistics management for fashion business	a. Develop new optimization heuristics to identify the solutions more efficiently b. Conduct analytical modeling research with models employing the advanced information technologies such as big data and cloud computing
10	Order-picking systems for fashion retailing companies	a. Extend the optimization models by relaxing some assumptions b. Applying the analytical models in other industrial settings
11	Inventory control with service level and lead time considerations	a. Include the pricing decision in the inventory control policy b. Explore channel coordination challenges c. Incorporate the stochastic component into the lead time formulation
12	Fashion sales forecasting by an analytic hierarchy process (AHP) scheme	a. Identify a method to aggregate quantitative variables properly so as to improve forecasting accuracy b. Test the time efficiency of the AHP scheme with the use of the much bigger data set

References

Bucci MJ, Woolard R, Joines J, Thoney K, King RE (2014) Incorporating economies of scale into facility location problems in carpet recycling. J Text Inst 105(12):1300–1311

Cachon GP, Swinney R (2011) The value of fast fashion: quick response, enhanced design, and strategic consumer behavior. Manage Sci 57(4):778–795

Caro F, Gallien J (2010) Inventory management of a fast-fashion retail network. Oper Res 58 (2):257–273

Caro F, Gallien J (2012) Clearance pricing optimization for a fast-fashion retailer. Oper Res 60 (6):1404–1422

Chiu CH, Choi TM, Tang CS (2011) Price, rebate, and returns supply contracts for coordinating supply chains with price dependent demands. Prod Oper Manag 20(1):81–91

Choi TM (2007) Pre-season stocking and pricing decisions for fashion retailers with multiple information updating. Int J Prod Econ 106:146–170

Choi TM (2013a) Optimal apparel supplier selection with forecast updates under carbon emission taxation scheme. Comput Oper Res 40:2646–2655

Choi TM (2013b) Optimal return service charging policy for fashion mass customization program. Serv Sci 5(1):56–68

Choi TM (2013c) Local sourcing and fashion quick response system: the impacts of carbon footprint tax. Transp Res Part E 55:43–54

Choi TM (2013d) Carbon footprint tax on fashion supply chain systems. Int J Adv Manuf Technol 68:835–847

Choi TM (2013e) Game theoretic analysis of a multi-period fashion supply chain with a risk averse retailer. Int J Inventory Res 2:63–81

Choi TM (2013f) Multi-period risk minimization purchasing models for fashion products with interest rate, budget, and profit target considerations. Ann Oper Res. doi:10.1007/s10479-013-1453-x (in press)

Choi TM, Chow PS, Xiao T (2012) Electronic price-testing scheme for fashion retailing with information updating. Int J Prod Econ 140:396–406

Choi TM, Hui CL, Liu N, Ng SF, Yu Y (2014) Fast fashion sales forecasting with limited data and time. Decis Support Syst 59:84–92

Donohue K (2000) Efficient supply contracts for fashion goods with forecast updating and two production modes. Manage Sci 46(11):1397–1411

Eppen GD, Iyer AV (1997a) Backup agreements in fashion buying—the value of upstream flexibility. Manage Sci 43(11):1469–1484

Eppen GD, Iyer AV (1997b) Improved fashion buying with Bayesian updates. Oper Res 45 (6):805–819

Hu ZH, Wei C, Yu XK (2015) Apparel distribution with uncertain try-on time by evolutionary algorithm. Int J Clothing Sci Technol 27(1):75–90

Hu ZH, Yu XK (2014) Optimization of fast-fashion apparel transshipment among retailers. Text Res J. doi:10.1177/0040517514538695 (published online)

Iyer AV, Bergen ME (1997) Quick response in manufacturer-retailer channel. Manage Sci 43 (4):559–570

Jia P, Govindan K, Choi TM, Rajendran S (2015) Supplier selection problems in fashion business operations with sustainability considerations. Sustainability 7:1603–1619

Li J, Choi TM, Cheng TCE (2014a) Mean-variance analysis of two-echelon fast fashion supply chains with returns contract. IEEE Trans Syst Man Cybern Syst 44:422–434

Li Y, Xu L, Choi TM, Govindan K (2014b) Optimal advance-selling strategy for fashionable products with opportunistic consumers returns. IEEE Trans Syst Man Cybern Syst 44:938–952

Petruzzi NC, Monahan GE (2003) Managing fashion goods inventories: dynamic recourse for retailers with outlet stores. IIE Trans 35:1033–1047

Shen B, Choi TM, Wang Y, Lo CKY (2013) The coordination of fashion supply chains with a risk averse supplier by the markdown money policy. IEEE Trans Syst Man Cybern Syst 43:266–276

Tereyagoglu N, Veeraraghavan S (2012) Selling to conspicuous consumers: pricing, production, and sourcing decisions. Manage Sci 58(12):2168–2189

Wang M, Liu J, Chan HL, Choi TM, Yue X (2015) Effects of carbon tariffs on duopoly market entry decisions and price competition: insights from textile firms of developing countries. Int J Prod Econ. doi:10.1016/j.ijpe.2015.07.025 (in press)

Wu AWD, Chiang DMH (2011) Fashion products with asymmetric sales horizons. Naval Res Logistics 58:490–506

Xiao T, Yang D, Choi TM, Cheng TCE (2014) Optimal reservation pricing strategy for a fashion supply chain with forecast update and asymmetric cost information. Int J Prod Res. doi:10.1080/00207543.2014.998789 (in press)

Yang D, Choi TM, Xiao T, Cheng TCE (2011) Coordinating a two-supplier and one-retailer supply chain with forecast updating. Automatica 47:1317–1329

Yeung WK, Choi TM, Cheng TCE (2010) Optimal scheduling in a single-supplier single-manufacturer supply chain with common due windows. IEEE Trans Autom Control 55:2767–2777

Yu Y, Choi TM, Hui CL, Ho TK (2011) A new and efficient intelligent collaboration scheme for fashion design. IEEE Trans Syst Man Cybern Part A 41:463–475

Zheng J, Chiu CH, Choi TM (2012) Optimal advertising and pricing strategies for luxury fashion brands with social influences. IEEE Trans Syst Man Cybern Part A 42(4):827–837

Chapter 2
Analytical Modeling Research Methodologies for Fashion Business Operations Management

Tsan-Ming Choi

Abstract In this paper, we examine the analytical modeling research methodologies for operations management in fashion business. We first concisely examine the related literature on analytical modeling research methods for operations management. Then, we propose the research frameworks for conducting analytical modeling operations analysis in fashion business. To be specific, by classifying the frameworks into economic modeling and engineering modeling, we propose the respective steps for conducting rigorous analytical modeling operations management studies in fashion business. For each step, some related important remarks specific to fashion business operations are discussed. Comparisons between these two categories of analytical modeling research methods are also presented.

Keywords Research method · Analytical modeling research · Mathematical models · Fashion business · Economic models · Engineering models · Optimization models

2.1 Introduction

Nowadays, operations management is a very important field of studies, and it is a spin-off from operations research and management science. In the fashion industry, operations management covers topics such as supply chain management, product design, quality control, production and capacity management, inventory control, sourcing, pricing and revenue management, and retail visual merchandising. Proper operations management for fashion businesses requires the support from scientific methods, in particular, the development of mathematical analytical models.

T.-M. Choi (✉)
Business Division, Institute of Textiles and Clothing, The Hong Kong Polytechnic University, Hung Hom, Kowloon, Hong Kong
e-mail: jason.choi@polyu.edu.hk

T.-M. Choi (ed.), *Analytical Modeling Research in Fashion Business*,
Springer Series in Fashion Business, DOI 10.1007/978-981-10-1014-9_2

In the literature, Ackoff (1956) is among the first group of operations researchers who develop a specific framework for conducting scientific analytical modeling research in operations. He proposes the well-known six-phase model which can systematically explore operations management problems ranging from inventory control, resource allocation, queueing, and routing, to replacement processes. Lathrop (1959) discusses science in operations research and comments that more scientific work in operations is necessary for the development of operations research and management science. Morris (1967) examines the art of analytical modeling research. He proposes three hypotheses on the proper approach in developing analytical models for management science studies. Among them, he believes that it is helpful to start with simple models even though they are quite distinct from reality and then move evolutionarily toward the more realistic models. This approach helps to avoid frustration which is very likely to appear when one considers a rich and complex model at the very beginning. Gass (1990) discusses the issues associated with analytical modeling research in which users act as modelers. He suggests that the proper model development methods and processes must be employed in order to yield high-quality insights and decision supports from the analytical models. Willemain (1994) interviews twelve modeling experts and reports some findings on the modeling practices and challenges. Many of the findings provide guidance and tips to proper analytical modeling for operations analysis. More recently, there are proposals on adopting multiple methodologies to examine operations management problems (Boyer and Swink 2008; Carter et al. 2008; Singhal and Singhal 2012a, b; Tang 2015). In particular, Van Mieghem (2013) examines the "3Rs," namely research, relevance, and rewards, of operations management studies. He argues that to have a sustainable future, operations management research must be relevant, both internally (to researchers) and externally (to practitioners). So he advocates the incorporation of rigorous empirical studies into analytical modelling to yield rewarding and fruitful research results in operations management. Echoing Van Mieghem's (2013) view, Simchi-Levi (2014) suggests ways to conduct operations management studies from a data-driven perspective. He proposes that real-world data-driven operations analysis presents a great opportunity and opens a new avenue to cultivate innovative and creative research in operations management. Gallien et al. (2015) explore the practice-based research in operations management. They argue that the real-world practice relevant operations management studies are driven by two dimensions of generalizability and validity. They discuss the related challenges and suggest ways to encourage more researchers to engage in practice-based operations management research. Most recently, Choi et al. (2016) examine the research trend of employing the multi-methodological approach in operations management. The authors discuss the strengths and weaknesses of the multi-methodological approach. They also present some specific multi-methodological approaches which can enhance the research rigor of operations management studies. As a remark, Choi et al. (2016) comment that the multi-methodological approach in operations management is in fact consistent with the classical framework proposed by Achoff (1956).

In this paper, based on the well-established analytical modeling research methods proposed by Achoff (1956), Morris (1967), and Gass (1990), we present how analytical modeling research can be conducted to explore fashion business operations management problems. We classify the fashion business operations management studies by dividing them into two different categories, namely the economic studies and the engineering studies. For each category, we introduce the respective steps of conducting analytical modeling research and show the related published research in the literature.

The rest of this paper is organized as follows. First, we present the economic analytical modeling framework, propose the respective steps, and review some related studies in Sect. 2.2. Then, we introduce the engineering analytical modeling research framework, explore the respective steps, and examine some related research in Sect. 2.3. We conclude this paper in Sect. 2.4.

2.2 Economic Models

In economics, an economic analytical model can be described as a theoretical construct expressed in mathematics, which is usually a simplified framework (compared to the real-world scenario), to demonstrate the complex economic processes. Operations management, as an important field of modern business studies, aims to generate important managerial insights by a scientific approach, and it requires the support of a solid methodology. The economic analytical modeling approach hence serves this need. As argued by many pioneering management scientists such as Morris (1967), it is wise and important to start the analysis by focusing on some simpler models. In particular, if we aim at deriving theoretically sound closed-form analytical tractable solutions for the operations management problems under studies, we usually have to employ simple models and impose some needed assumptions. With these points in mind, the steps for conducting economic modeling research in fashion business operations management are proposed as follows:

Step 1: *Formulating the problem*: Fashion business operations management is related to the fashion industry. It is thus of the upmost importance to identify the industrial needs. For example, is the problem related to some widely observed industrial problems and challenges in fashion business operations? These real-world industrial needs provide the needed "motivation" for conducting this piece of research. In addition, if high-quality (accurate and systematic) real datasets (e.g., demand data, inventory data, and lead time data) are available, we should carefully collect them and make use of them for the whole study. Similar to other kinds of business research, it is important to search the literature and conduct a comprehensive literature review on the related problems. Some related problems, models, research findings, and under-explored issues should be identified from the literature review. Then, we can formally define the scope of the problem (e.g., is it a supply chain inventory management issue? A dynamic pricing and revenue management issue? etc.) and proceed to Step 2.

Step 2: *Constructing the analytical model*: From Step 1, the specific research issues, the problem scope, and some related models should have been identified. In Step 2, we proceed to construct the specific analytical model by using standard mathematics. For example, we can construct an optimization model for the decision-making problem (e.g., on inventory management) and define the analytical objective function (e.g., maximizing the expected profit), the decision variable (e.g., the ordering quantity), and the constraints (e.g., on some realistic conditions such as budgeting, and the other model assumptions). As a remark, the proper way to establish the right economic analytical model is more art than science and there are no simple rules of thumb to help. However, some basic guidelines, as listed below, should be observed: 1. Start with a simple model and then proceed to refine it gradually with respect to the given data and the observed real-world practices. 2. Revise and adopt the model assumptions which are reasonably real world supported and can lead to analytical tractable results. 3. Choose the model understandable by the whole team of modelers. For more details, we refer readers to Morris (1967). Notice that in order to obtain analytically tractable closed-form solution, the economic analytical model developed in Step 2 is usually relatively simple. This point deserves much attention because most real-world fashion business operations are complex, while for conducting economic analytical modeling research for the respective operations problem, we have to confine ourselves to relatively simple models by setting the respective assumptions.

Step 3: *Deriving the solution*: After establishing the analytical model in Step 2, we need to identify the respective solution. Observe that the common methods for economic analytical modeling in operations management include optimization modeling and game theoretical modeling. Thus, in Step 3, if we have developed a standard optimization model in Step 2, our goal is to identify the optimal solution; if we have established a game theoretical model in Step 2, our goal is to determine the equilibrium solution. For economic operations management studies, we aim at deriving the closed-form solution. In many cases, we want to identify the unique solution (e.g., the unique equilibrium in the game theoretical analysis), which is especially helpful to derive managerial insights in Step 4. Notice that it is critically important to verify whether the solution makes sense by checking whether the solution is reasonable and in line with the reality (i.e., the practices observed in the fashion industry).

Step 4: *Generating managerial insights*: After determining the solution, we need to go deeper in generating insights from the solution. Standard studies, which include (i) the structural property analysis which aims to show the features of the solution (e.g., on solution uniqueness and special cases) and (ii) the sensitivity analysis which helps reveal how a change of model parameters affects the optimal solution (which also implies the rational decision of the related party) or the equilibrium, can be conducted. As a remark, the insights developed should be very practical and understandable by the fashion practitioners. As what Raman (1999) commented, managers in the fashion industry are not familiar with management science tools, and hence, we cannot expect they are able to appreciate the complex theoretical insights. They treasure simple and direct managerial advices.

Step 5: *Extending the model?* Since in Step 2, the analytical model being studied is relatively simple, one may extend it in Step 5 by considering a more general or complex situation. Usually, Step 5 involves relaxing some assumptions from the model explored in Step 2. Once the extended model is developed, similar analyses from Step 3 and Step 4 should be conducted. Of course, this proposed Step 5 is not necessary if the model developed in Step 2 is reasonably comprehensive, and the derived findings are sufficient. In the fashion industry, we have different kinds of operations (local versus global), different product types, and different kinds of product tiers (higher end noble brands versus lower end mass market brands). To extend the model, one can actually consider these differences and extend the analysis from one scenario to another scenario, with the goal of yielding generalizable managerial insights and theories.

Step 6: *Real-world validation*: Operations management in fashion should be related to the real industrial practices. As a scientific study, economic analytical modeling research should check the real-world situations and validate if the findings generated by the mathematical analysis are logical and consistent with real-world practices. This step on real-world validation has been highlighted in various classic studies and textbooks (e.g., Gass 1990; Taha 2002). We should also examine whether the findings can well explain real-world situations and phenomena. In fact, the goal of conducting economic analytical modeling research for fashion business operations management is to provide scientifically sound insights which can enhance the respective managerial practices. Empirical case studies, both qualitative and quantitative, industrial interviews, and surveys are all some relevant approaches to adopt in Step 6. Undoubtedly, in the fashion industry, the operations are usually rather complex. However, economic models are all, by definition, simplified versions of the real world. Thus, the findings based on the simplified economic models have to be validated with real-world practices in fashion in order to be convincing and scientifically solid.

Table 2.1 summarizes the steps involved in conducting economic analytical modeling research in fashion business operations management; some important remarks are also added. As shown in Table 2.1, even though economic analytical

Table 2.1 The steps included in the economic analytical modeling research framework

Steps	Key activities	Fashion business specific remarks
1. Formulating the problem	Literature review is critically important. Real-world practices and cases help to motivate the research	Fashion business operations management should be highly related to the fashion industry. Specific fashion company cases and data should hence be employed
2. Constructing the analytical model	Consider simple models first	Fashion business operations can be very complex, and it is wise to start from simple models
3. Deriving the solution	The existence of a unique closed-form solution is desirable	Verifying whether the solution makes sense to fashion is critically important
4. Generating managerial insights	The sensitivity analysis and the structural property analysis, both analytical and numerical, are commonly employed	The insights should be as practical as possible and should be understandable by practitioners in the fashion industry who are in general not experts in analytical modeling research
5. Extending the model?	Generalize the findings (e.g., relaxing assumptions) by exploring the extended model(s)	It is wise to extend the model by following the fashion business practices and industrial setting, from one scenario to another one
6. Real-world validation	Validate the research findings with respect to real-world practices	This is very critical as the findings derived from simplified economic models need not be applicable to the real industrial setting in fashion. This validation helps to make the research solid and scientifically sound

modeling research focuses on the mathematical analysis, real-world practices, data, and empirical industrial inputs are crucial in order to provide the complete picture of the whole fashion business operations management study.

In the literature, we can also find many examples employing the economic analytical modeling approach for fashion business operations management. For example, Iyer and Bergen (1997) study the quick response policy in a two-echelon apparel supply chain. The authors model the problem by considering a newsvendor type of product. Using the Bayesian conjugate pair demand distribution under the normal process, they reveal the impacts brought by quick response to the supply chain and its agents. In order to achieve Pareto improvement, they propose various measures such as service level commitment, volume commitment, and price commitment. Finally, the authors extend the model analysis to study the case with multiple items and also validate the findings by real-world industrial practices. Eppen and Iyer (1997) investigate the backup agreement in fashion buying. The authors construct the analytical model based on the observed industrial practice. They focus on exploring the value of upstream flexibility and generate important insights. By running a retrospective test of the model solution against the real buyer decisions, they find that backup agreements can enhance expected profits and may increase the amount of committed quantity. It is also encouraging to see that backup agreements may also benefit the manufacturer in maintaining its expected profit. Donohue (2000) examines supply contracting in a fashion supply chain with two production modes and forecast updating. The author considers the wholesale pricing with returns supply contract and proves that it can achieve supply chain coordination. In terms of theoretical contribution, her paper is among the first which explores supply chain coordination in the presence of forecast information updating. Jain and Paul (2001) study the operations reversal problem, which involves switching two consecutive stages in a manufacturing process for fashion products. The authors consider both customer heterogeneity and customer preference unpredictability in their model. They identify the analytical conditions in which operations reversal can lead to reduced production volume variability in the manufacturing process. Raman and Kim (2002) explore the impacts of inventory holding cost and the capacity management issue in apparel manufacturing. The authors focus on a real case study with a school uniform manufacturer. Based on many practices of this case, they construct the formal analytical models and then conduct analysis. Their findings reveal that apparel manufacturers with high inventory holding costs should set high stockout costs and achieve low capacity utilization. Choi (2007) studies the preseason inventory ordering and pricing problem in fashion retailing. The author makes several important industrial observations and establishes the stylish model. With multiple Bayesian information updating, he derives the ordering and pricing policy. By an extensive numerical sensitivity analysis, insights on the impacts brought by demand uncertainty and the value of information are generated. The author further extends the model with the consideration of different objective functions for determining the optimal retail price. Cachon and Swinney (2011) study the fast fashion supply chain with strategic consumers. The authors first observe the features of fast fashion supply chains and identify the core elements of them. Then, they explore how these features, such as quick response and enhanced design, relate to the operational performance of the fast fashion supply chain in the presence of forward-looking strategic consumers. They find that if a fast fashion company adopts both quick response and enhanced design strategies together, it will receive a higher profit compared to the sum of profits for the cases when it adopts either strategy alone. Chiu et al. (2011) study the supply chain coordination challenge with the use of "price, returns, and rebates" contract. The authors first report the industrial survey on the presence of the price, returns, and rebates supply contract in various companies, including many fashion-related brands. They then construct formal analytical models and explore the achievability of supply chain channel coordination by using the price, returns, and rebates contract when demand is price dependent and stochastic. Choi (2013) discusses the optimal product returns policy under the fashion mass customization program. Motivated by the observed mass customization industrial practices in various fashion brands such as Nike, the author builds a simple analytical model to explore the optimal consumer return service charging policy for fashion mass customization products. An extensive analytical sensitivity analysis is conducted to generate many important managerial insights. The author further extends the model

Table 2.2 The steps adopted by the related studies in economic analytical modeling research

Steps	Related studies									
	Iyer and Bergen (1997)	Eppen and Iyer (1997)	Donohue (2000)	Jain and Paul (2001)	Raman and Kim (2002)	Choi (2007)	Cachon and Swinney (2011)	Chiu et al. (2011)	Choi (2013)	Lee et al. (2015)
1	✓	✓	✓	✓	✓	✓	✓	✓	✓	✓
2	✓	✓	✓	✓	✓	✓	✓	✓	✓	✓
3	✓	✓	✓	✓	✓	✓	✓	✓	✓	✓
4	✓	✓	✓	✓	✓	✓	✓	✓	✓	✓
5	✓	✓	✓		✓	✓	✓		✓	✓
6	✓	✓			✓		✓		✓	

to the case when the fashion company possesses a risk-averse objective function. Finally, some real-world industrial practices are employed to verify the research findings derived from the analytical model. Most recently, Lee et al. (2015) study a supply chain with strategic and loss-averse consumers. The authors propose fast fashion inspired industrial measures such as quick response, enhanced design, and customized design strategies to enhance the agility and overall performance of the supply chain system. Table 2.2 lists the above reviewed studies in the scope of economic analytical modeling research related to fashion business operations management and shows the respective steps adopted in the analysis. As we can observe from Table 2.2, not all steps of our proposed framework are adopted in every cited reference. In particular, Step 6 seems to be overlooked by many studies.

2.3 Engineering Models

In Sect. 2.2, we have examined the economic analytical modeling approach for research in fashion business operations management studies. Even though real-world practices are considered in the economic modeling analysis, we usually do not expect to implement the model directly into any real-world application (e.g., a decision support system) because the objective is to identify the valuable managerial insights. However, if one aims to look at an operations management problem and identify an implementable solution scheme for real-world deployment (e.g., developing a software tool), then the engineering analytical modeling approach should be adopted.

Under the engineering analytical modeling approach for fashion business operations management, the focal point is on solving a realistic problem by incorporating as many real-world-related features and constraints into the model as possible. In other words, we plan to develop very realistic and "rich" optimization models which naturally would be highly complex. As a result, we do not intend to derive closed-form analytically tractable solutions; instead, we aim to identify the "best possible" solution for the realistic problem. Many methods such as the multi-criteria decision-making methods, the evolutionary algorithms, and the efficient heuristics would be applied to solve the related optimization problem.

Similar to the economic analytical modeling approach, we propose a six-step framework for the engineering analytical modeling approach as follows. Notice that the following framework follows the classic frameworks proposed by Ackoff (1956) and Hillier and Lieberman (1990).

Step 1: *Formulating the problem*: The goal is to identify the real problem from the respective fashion company and define it clearly. Thus, it is critically important to consider the related organizational behaviors in the fashion company, which can be learned by conducting extensive surveys and interviews with the people in the company. As a formal research study, it is also important to check the literature and identify related models and applications. Last but not least, data collection and management is very critical. In fact, for engineering analytical modeling research, data are the lifeblood as it directly drives the model development, the analysis, the testing, and also the implementation of the optimization tool. Poor data management will ruin everything. In the fashion industry, since not many companies are familiar with scientific operations management tools (see Raman 1999), we cannot take for granted that high-quality data are available. It is thus essentially important to collect the needed data and make sure the collected datasets are properly managed for model construction and further analytical studies.

Step 2: *Constructing the mathematical model*: This step is similar to Step 2 in the economic analytical modeling research we proposed above. However, the focal point here is to construct the mathematical model which is as realistic as possible. Having a complex optimization model is hence expected here. For many modelers, they would start establishing the model by considering a specific numerical instance of the observed operations problem. This would help them better understand the assumptions, constraints, and objectives and also let them have a taste of the problem complexity. As a remark, the completed analytical model in Step 2 should preferably be related to practices, supported by the literature, transparent (intuitive and understandable), robust, and scalable (can be extended and elaborated in the future when needs arise). Furthermore, it is important to seek advice from the practitioners in fashion for their comments on the model. This helps to enrich the model, facilitate its future implementation, and enhance the chance of success as well.

Step 3: *Deriving the solution*: Since the optimization model developed in Step 2 can be very complex, the solution may only be found numerically. In some cases, finding the globally optimal solution is computationally too expensive (e.g., for the NP hard problems) and hence one may be satisfied by finding the near-optimal or "satisfying solution." Heuristics and evolutionary algorithms are some widely applied approaches for solving realistic yet complex optimization problems. Notice that it is critically pertinent to verify whether the solution makes sense and is consistent with the fashion industrial practices (as many fashion companies are not using systematic measures in their operations practices).

Step 4: *Testing the model and solution*: Although we emphasize on the fact that the model we constructed in Step 2 is as realistic as possible, we have to understand that the model is "never more than a partial representation of reality" (Ackoff 1956). Thus, before deployment, we have to see how well the model and its solution perform. To do so, one can conduct a retrospective analysis (i.e., "creating the past"—using historical data to test the model and the solution) via, e.g., simulation studies. To be scientific, computational results from the simulation-based retrospective analysis should be verified by standard statistical tests (e.g., t test for comparing means). In Step 4, an essential task is to verify and validate the solution. Verification here refers to checking whether the solution scheme (e.g., the computerized algorithm) runs smoothly as planned. Validation means that whether the model and the derived solution are consistent with the reality. Observe that for the testing in Step 4, we should employ the real data collected from the fashion companies. The testing results, if positive, can hence provide convincing support to the model and the solution under tests.

Step 5: *Establishing controls:* In the optimization model for a real-world-related fashion business operations management problem, there are various model parameters. We have to be cautious on the fact that the solution of the optimization model is valid only when the model parameters are fixed like what we assume. However, in reality, the parameters may not be fixed or their values may deviate from our estimations. Thus, in Step 5, we have to establish control by taking some measures. For example, following the proposal by Ackoff (1956), we can do the following: For each parameter, we define the range of significant change. Then, we set up a sense-and-respond procedure to detect the occurrence of the significant change and provide reactive proposals on how the optimal solution should be revised if there is a significant change of a parameter. The sensitivity analysis will be helpful to identify the "significant change" in the optimization model. Note that similar to Step 4, the analysis conducted in Step 5 should be based on real data from fashion companies. In addition, as a lot of fashion companies have not established systematic data management and data quality auditing schemes, when we try to establish controls, we have to pay special attention to the issues associated with "data."

Step 6: *Implementing the solution*: After completing Steps 1–5, we proceed to explore the implementation and deployment of the solution. Special attention should be paid to the fact that the users of the solution are usually managers and business executives in the fashion company. They may not be analytically strong. Thus, it is important to make the tool easy to use and the solution easy to understand. Providing full documentation on the whole model, the implementation process and the related analyses can help, too. As a remark, some performance projections (e.g., on profit or cost) and performance metrics (for future checking) on adopting the proposed solution should also be provided.

Table 2.3 summarizes the steps of the engineering analytical modeling research framework for exploring fashion business operations problems and some related useful remarks. As we can see from the proposed steps, real-world practices, realistic models, and solution implementations are all focal points in the engineering analytical modeling research framework.

In the literature, there are many studies employing the engineering analytical modeling research framework. For instance, Fisher et al. (2001) explore the optimal inventory replenishment policy for fashion retailing. The authors formulate the

Table 2.3 The steps included in the economic analytical modeling research framework

Steps	Key activities	Fashion business specific remarks
1. Formulating the problem	Focus on the real company problem and scenario	Data collection and data quality control are most critical because they directly affect the model construction. In addition, expectedly, the final solution will be implemented to enhance practice, which also requires the support by using data
2. Constructing the analytical model	Consider the most realistic optimization model. Model verification and validation are essentially important	Get comments from fashion practitioners if possible. This helps to enrich the model and enhance the feasibility of having an applicable tool and solution
3. Deriving the solution	Find the efficient way of identifying the best possible solution	Verifying whether the solution is consistent to the fashion industrial practices and fashion practitioners is critically important
4. Testing the model and solution	Conduct the retrospective test	The tests should be based on real data and the results should be understandable by fashion practitioners
5. Establishing controls	Identify and define the significant change and impose procedure to detect and respond to it	This requires the real industrial data from fashion business operations. Since many fashion companies may not have very systematic data management and quality control, special attention should be paid to the "data"
6. Implementing the solution	Pay special attention to users	It is important to well document the whole model, the implementation process, and the related analyses and details so that fashion practitioners can trace back and check whether problems arise

problem as a two-stage stochastic dynamic programming optimization model and develop an important heuristic to solve it. They derive the conditions under which the proposed heuristic can efficiently determine the optimal solution. They apply the heuristic in a catalog retailer case and find that their proposed procedure significantly outperforms the existing practice. They further comment that their proposed method can be used to identify the optimal reordering time and select the optimal replenishment contract. Motivated by the observed challenges in the knitted fabric production process, Laoboonlur et al. (2006) study the respective production scheduling problem. They focus on a specific kind of knitted fabric dyeing and finishing process. By building the formal optimization model, they solve the problem and discuss the respective applications. Caro and Gallien (2010) study the inventory management challenges of the fast fashion retailer "Zara." The whole study is based on Zara's industrial practices and operations challenges. Realistic models are constructed. The authors report the real-world implementation of their proposed inventory management model and the respective inventory allocation process. Their experiments reveal that the new inventory allocation process can yield a good improvement of sales, increase the proportion of time of product display, and reduce transshipment in Zara. Yeung et al. (2010) investigate the optimal scheduling problem in a single-upstream-supplier and single-downstream-manufacturer supply chain system. Their model and optimization problem are inspired by various observed industrial practices in apparel production in China. After constructing the optimization model and uncovering the structural properties, an efficient algorithm is developed to help identify the optimal production schedule. Real datasets from an apparel manufacturer are also employed to further verify the performance of the proposed algorithm. Other recent studies employing the engineering analytical modeling research approach include the following: the study on clearance pricing in the fast fashion company Zara (Caro and Gallien 2012), the facility location planning problem for carpet recycling operations (Bucci et al. 2014), the fast fashion sales forecasting decision supporting systems (Choi et al. 2014), and the panel data-based fashion sales forecasting models (Ren et al. 2015). Table 2.4 lists the above reviewed papers and indicates the corresponding adopted steps (with respect to our proposed engineering analytical modeling research

Table 2.4 The steps adopted by the related studies in economic analytical modeling research

Steps	Related studies							
	Fisher et al. (2001)	Laoboonlur et al. (2006)	Caro and Gallien (2010)	Yeung et al. (2010)	Caro and Gallien (2012)	Bucci et al. (2014)	Choi et al. (2014)	Ren et al. (2015)
1	✓	✓	✓	✓	✓	✓	✓	✓
2	✓	✓	✓	✓	✓	✓	✓	✓
3	✓	✓	✓	✓	✓	✓	✓	✓
4	✓	✓	✓	✓	✓	✓	✓	✓
5	✓	✓	✓	✓	✓	✓	✓	✓
6	✓	✓	✓		✓	✓		

framework). Similar to the case in the economic analytical modeling research, Table 2.4 shows that not all proposed steps are adopted in each cited study. In particular, as many of these studies are academic in nature, Step 6 on real-world implementation is not completed in many of them.

2.4 Concluding Remarks

We have explored in this paper the research methodology for conducting analytical modeling research in fashion business operations management. We have reviewed the extant literature on analytical modeling research in operations management. We have proposed the research frameworks for two categories of analytical modeling research, namely the economic analytical modeling research and the engineering analytical modeling research. Each proposed research framework includes six steps which provide guidance to researchers and practitioners on the proper way of conducting operations management research in fashion. For each step of the research framework, some relevant remarks specific to fashion business operations are discussed. As a concluding remark, we further present in Table 2.5 the comparisons between the economic analytical modeling research and the engineering analytical modeling research frameworks.

From Table 2.5, we can see that both frameworks are systematic (with six steps) and have a focal point on being scientifically sound. However, they differ in terms of the research goal, model simplicity, solution features, and the popular tools employed for the analysis. Before closing, it is important to understand that the division of fashion business operations management analytical modeling research into "economic models" and "engineering models" is just a way to systematically classify the related works and present the respective research frameworks. However, many related studies may fall in the middle between these two categories

Table 2.5 Comparisons between the economic analytical modeling research and the engineering analytical modeling research frameworks

	Economic analytical modeling research	Engineering analytical modeling research
Goal	Generate valuable managerial insights	Develop applicable models and solutions
Model simplicity	Simple models	Realistic models which are usually complex
Solution features	The globally optimal closed-form solution	The best possible solution
Process	Systematic, with 6 steps	Systematic, with 6 Steps
Scientifically sound?	Yes	Yes
Popular tools	Game theoretic analysis, global optimization	Simulations, evolutionary algorithms, heuristics

of models. Thus, this classification does not imply that the world of fashion business operations management analytical modeling research must be purely "black-and-white," and we have to understand that "gray" does exist.

For future research, it is meaningful to compare the steps of the two proposed analytical modeling research methodologies and generate deeper insights. One may also explore whether the two proposed analytical modeling research methodologies are comprehensive enough. Very likely, some further enhancement is possible. For example, for fashion business operations management studies which incorporate human behaviors into the analysis (Croson et al. 2013), the respective analytical modeling research methodology will probably be different from the ones discussed in this paper. In addition, under the trendy proposal of conducting multi-methodological research in operations management (Choi et al. 2016), it is also interesting to examine and establish the corresponding frameworks for fashion business operations management.

Acknowledgements The author sincerely thanks Bin Shen and Ying Zhang for their comments on the earlier version of this paper.

References

Ackoff RL (1956) The development of operations research as a science. Oper Res 4(3):265–295

Boyer KK, Swink M (2008) Empirical elephants—why multiple methods are essential to quality research in operations and supply chain management. J Oper Manage 26:338–344

Bucci MJ, Woolard R, Joines J, Thoney K, King RE (2014) Incorporating economies of scale into facility location problems in carpet recycling. J Text Inst 105(12):1300–1311

Cachon GP, Swinney R (2011) The value of fast fashion: quick response, enhanced design, and strategic consumer behavior. Manage Sci 57(4):778–795

Caro F, Gallien J (2010) Inventory management of a fast-fashion retail network. Oper Res 58 (2):257–273

Caro F, Gallien J (2012) Clearance pricing optimization for a fast-fashion retailer. Oper Res 60 (6):1404–1422

Carter CR, Sanders NR, Dong Y (2008) Paradigms, revolutions, and tipping points: the need for using multiple methodologies within the field of supply chain management. Journal of Operations Management 26:693–696

Chiu CH, Choi TM, Tang CS (2011) Price, rebate, and returns supply contracts for coordinating supply chains with price dependent demands. Prod Oper Manage 20(1):81–91

Choi TM (2007) Pre-season stocking and pricing decisions for fashion retailers with multiple information updating. Int J Prod Econ 106:146–170

Choi TM (2013) Optimal return service charging policy for fashion mass customization program. Serv Sci 5(1):56–68

Choi TM, Cheng TCE, Zhao X (2016) Multi-methodological research in operations management. Prod Oper Manage 25(3):379–389

Choi TM, Hui CL, Liu N, Ng SF, Yu Y (2014) Fast fashion sales forecasting with limited data and time. Decis Support Syst 59:84–92

Croson R, Schultz K, Siemsen E, Yeo ML (2013) Behavioral operations: the state of the field. J Oper Manage 31:1–5

Donohue K (2000) Efficient supply contracts for fashion goods with forecast updating and two production modes. Manage Sci 46(11):1397–1411

Eppen GD, Iyer AV (1997) Backup agreements in fashion buying—the value of upstream flexibility. Manage Sci 43(11):1469–1484
Fisher M, Rajaram K, Raman A (2001) Optimizing inventory replenishment of retail fashion products. Manuf Serv Oper Manage 3(3):230–241
Gass SI (1990) Model world: danger, beware the user as modeler. Interfaces 20(3):60–64
Gallien J, Graves SC, Scheller-Wolf A (2015) Practice-based research in operations management: what it is, why do it, related challenges, and how to overcome them. Manuf Serv Oper Manage (published online)
Hillier FS, Lieberman GJ (1990) Introduction to operations research, 5th edn. McGraw-Hill, New York
Iyer AV, Bergen ME (1997) Quick response in manufacturer-retailer channel. Manage Sci 43 (4):559–570
Jain N, Paul A (2001) A generalized model of operations reversal for fashion goods. Manage Sci 47(4):595–600
Laoboonlur P, Hodgson TJ, Thoney KA (2006) Production scheduling in a knitted fabric dyeing and finishing process. J Text Inst 97(5):391–399
Lathrop JB (1959) Operations research looks to science. Oper Res 7(4):423–429
Lee CH, Choi TM, Cheng TCE (2015) Selling to strategic and loss-averse consumers: stocking, procurement, and product design policies. Naval Res Logistics 62:435–453
Morris WT (1967) On the art of modeling. Manage Sci 13(12):B707–B717
Raman A (1999) Managing inventory for fashion products. In: Tayur S, Ganeshan R, Magazine M (eds) Quantitative models for supply chain management. Kluwer, Massachusetts, USA, pp 790–805
Raman A, Kim B (2002) Quantifying the impact of inventory holding cost and reactive capacity on an apparel manufacturer's profitability. Prod Oper Manage 11(3):358–373
Ren S, Choi TM, Liu N (2015) Fashion sales forecasting with a panel data based particle filter model. IEEE Trans Syst Man Cybern Syst 45:411–421
Simchi-Levi D (2014) OM research: from problem-driven to data-driven research. Manuf Serv Oper Manage 16(1):2–10
Singhal K, Singhal J (2012a) Imperatives of the science of operations and supply-chain management. J Oper Manage 30:237–244
Singhal K, Singhal J (2012b) Opportunities for developing the science of operations and supply-chain management. J Oper Manage 30:245–252
Taha HA (2002) Operations research: an introduction, 7th edn. Prentice Hall, NJ
Tang CS (2015) Making OM research more relevant: "why?" and "how?" Manuf Serv Oper Manage (in press)
Van Mieghem JA (2013) Three Rs of operations management: research, relevance, and rewards. Manuf Serv Oper Manage 15(1):2–5
Willemain TR (1994) Insights on modeling from a dozen experts. Oper Res 42(2):213–222
Yeung WK, Choi TM, Cheng TCE (2010) Optimal scheduling in a single-supplier single-manufacturer supply chain with common due windows. IEEE Trans Autom Control 55:2767–2777

Chapter 3
A Review: Customer Returns in Fashion Retailing

Mack A. Diggins, Chen Chen and Jing Chen

Abstract Customer returns are a common practice in the retail industry. Due to the unique characteristics of fashion retailing, including short product life cycle, high returns rates, fraudulent returns, and low salvage value, customer returns are critical to fashion retailers. In this paper, we briefly discuss issues related to customer returns in fashion retailing. We then review the studies in the literature considering the impact of customer returns policies, methods of reducing customer returns, and management decisions related to customer returns, as these topics relate to or can be applied to the fashion retailing industry. We finish with a discussion of future research directions on managing customer returns in the fashion retail industry.

Keywords Customer returns · Customer returns policy · Fraudulent returns · Fashion retailing · Fashion market · Literature review

3.1 Introduction

Returns of products from customers to retailers are a common phenomenon in the retail industry. Some consumers return products that perform unsatisfactorily, while others return products that function satisfactorily for other reasons, such as not meeting expectations or tastes. Returns are growing at a significant rate; Stalk (2006) reported that "returned goods are estimated to exceed $100 billion per year in the United States and in many categories; the number of returns is growing at around 50 % a year". Customers receive either a partial or full refund for the

M.A. Diggins (✉) · C. Chen · J. Chen
Rowe School of Business, Dalhousie University, 6100 University Avenue, Halifax, NS B3H 4R2, Canada
e-mail: mack.diggins@dal.ca

C. Chen
e-mail: c.jesschen@gmail.com

J. Chen
e-mail: jchen@dal.ca

T.-M. Choi (ed.), *Analytical Modeling Research in Fashion Business*, Springer Series in Fashion Business, DOI 10.1007/978-981-10-1014-9_3

returned product from a retailer, according to specific returns policy articulated by the retailer. The National Retail Federation (NRF) reported in "2014 Consumer Returns in the Retail Industry" that total merchandise returns accounted for almost $284 billion in lost sales for US retailers.

Toktay (2003) stated that customer return rates ranged from 5 to 9 % of sales for most retailers. Returns rates, however, vary greatly between industries (Langley et al. 2008), categories (Barry 2000), and distribution channels (Ofek et al. 2011). Barry (2000) pointed out that "Not surprisingly, the category with the highest rate of return is fashion apparel, at a whopping 25–40 %". In comparison, business products are rarely returned. It is easy to see that items that involve size, colour, fashion, or that unquantifiable "it's not me" factor, are more likely to be returned (Barry 2000). While typical returns rates in a traditional retail channel average at 8.9 % (according to the NRF's report, 2014), retailers in the USA report a return rate of between 20 % and 40 % for online sales, with poor fit cited as the number one reason (Ratcliff 2014). A recently released study by Murphy and Berns (2015) with HRC Advisory also suggests that online returns are expensive and present an important barrier for retailers adopting a customer-centric, omni-channel model.

Although free shipping and lenient return policies have given online retailing a huge boost (Banjo 2013), online retailers, especially online fashion retailers, may experience serious customer returns problems. Returns rates for online sales can be much higher than in a brick-and-mortar channel (Kim 2013; Millar 2013). Since "fast fashion" is one of the main characteristics of the fashion industry throughout the world (Thompson 2013), and consumers are progressively spending more time online, the rise of mobile and e-commerce is especially relevant in the fashion industry (Cecilio 2015). For apparel items, return rates are around 20 % for brick-and-mortar retailers and twice that for online sales (VandeVate and Bedir 2005; Martinez 2009). As pointed by Banjo (2013), behind the uptick in e-commerce, however, is a little known secret that as much as a third of all Internet sales gets returned. European online retailers experience product returns rates of 40 % or higher in product categories such as fashion. Mostard and Teunter (2006) stated that return rates of certain fashion items have been estimated to be as high as 74 %.

The increase of online sales in fashion retailing creates new challenges, as customers cannot try on clothes prior to purchase, and higher return rates result (see Walsh et al. 2014). Customer returns is positively correlated with losses in sales, as well as with significant costs associated with handling returned products at the retailer. The Economist (2013) showed that handling each returned item costs an online retailer between $6 and $18, in arranging staff, receiving, repacking, and restoring the returned products (see also Banjo 2013). Costs are not the only consideration; returns demand precise reverse logistics as well as excellent customer service is needed to ensure that the customer is satisfied and will be likely to return and shop with the same retailer (Melendez 2013).

As competition intensifies, retailers are forced to create sustainable competitive advantages while adding value for their customers. One way to do this is through product returns management: offering a customer returns policy and managing the reverse logistics.

The costs of reverse logistics vary. Grant et al. (2006) concluded that the cost of moving a product back from the consumer to the producer could be as much as nine times that of the forward flow. Each return has to be opened, diagnosed, and then processed (Norek 2002). Furthermore, product returns may diminish current assets because of lower inventory values of returned products, increase order cycle times due to the reshipment of ordered items, increase short-term liabilities due to required repairs and refurbishments, and reduce revenues due to lost sales (Min and Ko 2008). Hewitt (2008) suggested that returns can reduce profits by 30–35 %. Consequently, it has become important to predict and control the costs of product returns by forecasting how much will be returned from which customer and why (Foscht et al. 2013). The findings of Stuart et al. (2005) demonstrated how the returns process can be re-designed to reduce costs and improve order fulfilments. Their research suggested that in order to make the algorithm more efficient, the re-design needs to consider backorders earlier in the inspection stage, as well as inventory level, net return value, lead time, and marketing patterns.

Gecker and Vigoroso (2006) stated that many companies identify the management of product returns as the most challenging aspect of reverse logistics. Chen and Bell (2009) pointed out that product returns represent a serious problem for both retailers, who must process the incoming returns, and the product suppliers, who must cope with the inventory of returned products. In addition, suppliers and retailers must negotiate some form of agreement around how physical returns are handled and who pays the costs of processing the returns and refunding customers. These decisions include, among other things: what returns policy should be offered (money-back guarantee, partial refund, or no refund), price for the product and wholesale price, refund price for the returned product, and order quantity.

Although the issue of customer returns has attracted ongoing attention in academic studies, studies focussed on fashion retailing are very limited. This paper examines the current literature on customer return-related decisions and practice in the fashion retailing industry. In particular, it reviews factors and approaches that are unique to customer returns within fashion retailing, the current state of academic research in managing customer returns, and its applications in fashion retailing, and finishes by suggesting some possible directions for future research.

3.2 Issues of Customer Returns in Fashion Retailing

One reason why fashion retailing faces a high returns rate is that many fashion retailers sell fashion products through Web sales and catalogues and offer lenient return policies. Web sales cannot provide the customers with "feel and touch" experience of the product to determine how well the product will fit their tastes and expectations (Ofek et al. 2011). It is difficult to judge the appropriateness of a product by reviewing the description and displays on a Website. Fashion products are generally categorized as high-risk purchases because of the sensory and interactive nature of the purchase process (Bhatnagar et al. 2000). Physical examination of fashion products to assess the fabric, colour, size, design, fit, and match with other items could be crucial for customers in their purchasing decisions (Ha and Stoel 2004). Due to the absence of "feel and touch" experiences, ordering decisions for fashion products are deemed to carry more risk. In order to reduce customers' risk, retailers offer lenient return policies (Wood 2001), although there is no law that requires any retail merchant to have a specific refund policy. Generous return policies ("no questions asked") result in significant customer returns rates, which can be reduced through restrictions on returns (such as asking for the return of the original packaging materials or requiring no visible signs of use) (Davis et al. 1998). Restrictive policies can range from offering no returns at all, to offering shorter time limits, requiring restocking fees, requiring a receipt, or requiring the original packaging (Kang and Johnson 2008). As pointed out by Shulman et al. (2009) charging restocking fees can discourage customers from returning the products.

Return policies are important in fashion retailing because they affect customer purchase behaviours and decisions on whether or not to return the product the customers have purchased. Fashion retailers often offer money-back guarantees or allow refunds for any reason ("no questions asked" policy), even if the product or service had adequately fulfilled its intended function (Che 1996; Davis et al. 1995, 1998). For instance, Zappos (an online fashion retailer specializing in selling shoes, accessories and clothing) has proven to see results in using a lenient return policy. Zappos proudly advertises their free delivery and returns for fashion products purchased online. As a result, it has built and retained their customer loyalty. Its Webpage also stresses the leniency of their returns policy by highlighting that customers are able to return items for any reason and can even return products one year after the product had been purchased online. If customers then decide they want to return an item, Zappos provides customers with the ability to print the mailing label and return the item to the nearest post office (Zappos 2015). To reduce higher returns rates as well as fraud returns, most fashion brands provide a full-refund policy for products in the original condition, such as in original packaging, unworn, and with original labels. Some allow full refund with some restrictions. For instance, Nike's return policy requires, such as "must return product within 30 days of the purchase date", "products must be unworn, unwashed, and in the original packaging", and "provide a valid receipt and photo ID". Furthermore, when customers meet the guidelines for returning a product, if an item is purchased in store, it must be returned at the same location (Nike 2015). Table 3.1 summarizes consumer returns policies of some brands of fashion product.

Consequently, return policies are an aspect of marketing practice (Autry 2005), and therefore, returns management is certainly an aspect of the value creation process (Hjort et al. 2013). Lee (2009) pointed out fashion products purchased at home via online shopping or catalogue shopping are much more likely to be returned than any other types of products because such fashion products can easily be returned and products that have been tried on or damaged are often returned.

Fashion retailing is typically characterized under the consumer goods market, which is described by short life cycles; high volatility; low predictability; fickle

Table 3.1 Returns policies for some brands of fashion product

Laura	Returns: within 14 days; exchange: after 14 days
Stylebop.com	Within 14 days of purchase
Ally fashion	Within 14 days of delivery
Simons	30 days for regular-priced items and 10 days for sale items; shipping fee is non-fundable
Dynamite	Within 30 days
John Lewis	Within 90 days

The findings are based on our online searching of the reported consumer returns policies in these fashion brands

consumer preferences; high impulse purchasing; and a myriad of manufacturing, marketing, and retail alternatives (Richardson 1996; Christopher et al. 2004; Masson et al. 2007; Fernie et al. 2010). Fashion products, along with products such as perishables and high tech products, are initially very interesting for the customer, but as time passes, these products become less popular (Ahangarkolaei and Zandi 2010). Thus, when these products are returned, they have very low salvage value, and this makes product returns management even more challenging than in other markets. Some aspects of customer returns in the fashion industry, however, are unique.

Some customers may take advantage of generous return policies and buy merchandise with the intent to return it once it has been used for a specific purpose (Piron and Young 2001). With lenient return policies, consumers are also more likely to commit return fraud (Bahn and Boyd 2014). As classified by Foscht et al. (2013), this behaviour is referred to variously as a "moral hazard in consumption" (Davis et al. 1998), "retail borrowing" (Piron and Young 2001), "deshopping" (King 2004; King et al. 2008), or "fraudulent returning" (Zabriskie 1973; Harris 2008, 2010). Schmidt and Sturrock (1999), as well as King et al. (2007), reported that this behaviour is a growing trend in consumer shopping behaviours. A report from the National Retail Federation showed that return fraud and abuse is about 6.1 % of total returns in US retailing in 2014. Piron and Young (2000) found that 18 % of customers they investigated had experience of product borrowing, while a majority (98 %) of the borrowed products were apparel items. Social need (e.g. to attend a wedding) was the main reason for fashion product "borrowing". Specifically, customers who carry out product "borrowing" often return high-priced suits or dresses immediately after they use them for special occasions (Piron and Young 2000). Through interviews and data analysis; Harris (2008) identified ten main factors that appear to be related to customers' likelihood of fraudulently returning products. The study provided insights to practitioners who routinely deal with incidents of fraudulent returns.

Handling of the returned products becomes a serious issue in fashion retailing. From a survey, Creutz and Larsson (2012) pointed out how the fashion industry is faced with greater difficulties when handling the needs in the supply chain and reverse logistics processes, since the industry is challenged with shorter life cycles and an unpredictable market.

Another special case is mass customization. As innovations in production technology and efficient operations continue apace, many products can be produced through mass customization of production processes. As pointed out by Choi et al. (2013) and Choi (2013), it is common that a no-refund policy will be applied to fashion products offering customization (such as monogramming). It may be less easy to decide what returns policy should be adopted for products created using mass customization production processes.

Finally, in the fashion e-commerce business, a trend towards more liberalized delivery and return conditions has become evident. This trend represents a method of coping with competition inside the industry while simultaneously attracting new consumers from the traditional retail chains.

3.3 Managing Customer Returns

In this section, we will review literature on managing customer returns: the impact of returns policies, handling customer returns, and decisions in the presence of customer returns, with a focus on fashion retailing. As we mentioned above, customer returns-related issues have received extensive studies in the past two decades, but few studies have focused on customer returns-related issues in fashion retailing. We thus propose some research directions for further research.

3.3.1 The Impact of Customer Returns Policies

The core concept of marketing is to drive maximal profit by employing service strategies to satisfy customers. Service strategies include better shopping environment and experiences (Boyaci and Gallego 2004), better product availability and on-time delivery (Chen et al. 2008), and post-purchase services (Lele 1997; Cohen et al. 2000). Among the variety of services provided by retailers, the ability to accommodate customer returns of unsatisfied products represents one of the key competitive advantages. Davis et al. (1998) analysed the return policies of 133 retailers. They showed that return policies vary with how quickly a product is consumed, the salvage value of returned merchandise, and whether there are opportunities to cross-sell or substitute other items when returns occur. Davis et al. (1995) showed that fashion retailers are more likely to accept returns of “regularly priced merchandise” than “clearance” items; this provides evidence that return policies are more liberal when the product has a higher salvage value. Xu et al. (2015) pointed out that in general, retailers give a short allowable return period to products with a short life cycle, such as seasonal products and fashion products, while give a relative long allowable return period to products, such as durable goods.

Stalk (2006) studied a case of a fashion retailer, Baby Gap and Gap Kids, to provide an example of the importance of providing a consumer-friendly returns policy. His research outcome suggested that companies should look at return policies as a way to market their brand and increase the level of customer loyalty towards the brand. He also suggested that offering a generous and easy return policy would attract more customers and it could be used as a strategy to gain competitive advantage. Pralle and Stalk (2006) also explained using return policies as a marketing tool to attract more customers. Petersen and Kumar (2010) analysed six-year purchase data and concluded that a lenient return policy can encourage customers to be more willing to make other purchases, thereby increasing the company's revenues from increased sales.

From the perspective of customers, Bahn and Boyd (2014) concluded that implementing a restrictive return policy can be detrimental towards a customer's behaviour. They reasoned that potential consumers who are faced with a restrictive return policy may experience high levels of perceived risk. Biederman (2005) explained that online shoppers would not revisit a retailer who did not provide a returns policy. Bonifield et al. (2010) found that there was a positive relationship between the level of leniency of return policies and consumers' perception of quality in an online retailer. Desmet (2014) demonstrated how a money-back guarantee reduces the perceived risk a consumer might feel towards a product, implying that charging a premium price (required to cover the guarantee) signals greater quality. Ramanathan (2011) showed that performance of e-commerce in terms of product returns will have a significant impact on customer loyalty, which depends on the risk characteristics of the product. For instance, a returns policy will be of greater significant to customer loyalty. Che (1996) indicated factors to determine when it is beneficial to offer a return policy, including a consumer's degree of risk adversity, whether the product is considered an experienced good and retail price. Pei et al. (2014) empirically examined the effect of a return policy (the degree of leniency) on a consumer's perceived fairness of a return policy and purchase intentions using a structural equation model (SEM).

3.3.2 *Reducing Customer Returns*

The second stream of research focuses on how to reduce customer returns. Barry (2000) noticed that returns are increasing, in particular with products involving: size, colour, and fashion style. He provided a three-step process in order to reduce the amount of returns: measuring and analysing the causes of the returns, calculating costs associated with returns policies, and developing a strategic plan to reduce the costs while simultaneously maintaining a high level of customer service. Anderson et al. (2006) demonstrated how return rates can be better predicted by understanding that retail price and return rates are positively correlated. Furthermore, after analysing historical returns of an online fashion retailer, they showed that customers are more likely to return products if more sizes are available, but availability of more colours has potential in decreasing the likelihood of the products being returned. (Foscht et al. 2013) found through their research that perceived uncertainty with risk in mail orders of fashion products increases the probability of a consumer returning a product. Online shoppers may order multiple sizes and colours as a risk alleviation strategy, in order to have the ability to return the products that did not match their criteria. The research also found that people in similar demographics tend to shop similarly. Therefore, retailers can adjust a return policy by identifying "returning groups". Heiman et al. (2001) found that retailers can use money-back guarantees and demonstrations to reduce non-defective returns, in order to compensate for a lack of information and knowledge at the time of purchase. Explaining how an online retailer can turn an unpleasant product experience into a pleasant service experience, Ramanathan (2011) discussed how providing an easy and hassle-free returns process will increase the customer's experience and trust for the retailers' Website and increase the likelihood that the customer will repurchase from the same online retailer. Chu et al. (1998) investigated how an optimal partial refund rate (what customers pay to have the right to return items) can be determined in order to reduce the frequency of fraudulent returns. Hess et al. (1996) proposed using partial refunds to curb returns in online sales.

Mass customization has become a competitive strategy that a lot of fashion retailers have adopted. However, there are difficulties when dealing with returns and the associated charges for these mass customization products. Hjort and Hellstrom (2014) explained how the impact of delivery and return policies has an effect on consumer behaviour. They showed and explained how offering free delivery and free returns significantly increases the quantity of apparel sold. When an online fashion retailer offers a free return policy, this tends to decrease the average value of products being returned. Moreover, the research suggested that in order to optimize the effects of offering a free return policy on consumer shopping behaviours, the fashion e-commerce company must also provide a free delivery policy.

A frequent reason for apparel returns is improper fit. Online shoppers cannot try on garments before they purchase, and this increases the likelihood of poor fit, which increases the likelihood of customer returns. The patent developed by Rose (1996) provided innovative software to facilitate the experience of electronic fashion shopping. This software will help customers to choose the right fit and style to match their body type.

Bernon et al. (2011) adopted a case study to explore supply chain integration (SCI) practices and their implications in a retail product returns process between an OEM and its two retailers. They found that reducing retail product returns could significantly benefit both an OEM and its customers when appropriate SCI enabling practices are deployed. Ferguson et al. (2006) reported on the cooperation between manufacturers and retailers who reduced non-defective returns by training their sales force to communicate well with customers or by improving the description of the products. They investigated retailers' efforts to reduce returns of non-defective products in order to get rewards from the manufacturer (through a target rebate contract) if returns could be lowered to a certain level.

3.3.3 Decisions Related to Customer Returns

Studies on decisions related to customer returns discuss extensively the choice of MBGs or no-refund policies by comparing the payoffs from the two different policies, and deciding whether or not to implement a partial refund policy with a determination of refund prices.

Alptekinoglu and Grasas (2014) investigated how a retailer's product assortment decision is influenced by its associated return policy. They assumed that the retailer selects its assortment from an exogenous set of horizontally differentiated products. Consumers will then make a purchase and decide to keep/return based on the nested multinomial logit fashion. They found that the optimal assortment has a distinct structure for relatively strict return policies. Li et al. (2013) investigated the impact of an online retailer's return policy, product quality, and pricing strategy, and its effects on the customer's purchasing and return decisions, by developing theoretical models.

MBGs (or full-refund policies) are widely used in practice as customers perceive low risk when purchasing a product with an advertised full-refund policy. MBGs can enhance the store's image (Davis et al. 1995), increase customers' loyalty (Schmidt and Kernan 1985), increase storewide profits (Schmidt and Kernan 1985), indirectly stimulate growth in sales (Davis et al. 1995), generate customers attentions (Anderson et al. 2006), reflect high product quality (Moorthy and Srinivasan 1995; Shieh 1996), reduce consumers' perceived risk (Grewal et al. 2003; Lei et al. 2008), increase consumer satisfaction (e.g. McCollough and Gremler 2004), enhance purchase intentions (Davis et al. 1995; Wood 2001), and help retailers reduce costs on in-store demonstrations (Davis 1995). MBGs tend to protect customers against product mismatch. However, offering a hassle-free full-refund policy can impose substantial costs on retailers.

In the literature, there are two types of models to describe demand: using utility function to derive demand and using aggregate demand assumption.

Most of the studies on MBGs have, until recently, considered a monopolistic retailer. With survey data, Davis et al. (1998) argued that a MBG with some degree of hassle or inconvenience is better than a partial refund policy, because it is likely to be less negatively perceived by consumers. They assume that the consumer's potential value from the product is a random variable. Consumers know the distribution of the valuation, but no individual knows the exact value she will receive until the product is bought and tested. Developing a monopoly model, they showed that a retailer is more likely to offer a low-hassle return policy when: (1) its products' benefits cannot be consumed in a short period of time; (2) its product line offers opportunities for cross-selling; and (3) it can obtain a high salvage value for returned merchandise. Davis et al. (1995) studied a monopolistic retailer facing risk-neutral consumers who are equally satisfied with the product but are different in product valuation. By comparing optimal retailer profits under MBG and no MBG, respectively, they concluded that the retailer should offer MBGs if it can handle customer returns efficiently, that is, the salvage value of the returned product more than offsets the total transferring costs from the retailer (handling cost) and consumers (cost incurred when they return the product) related to the product return.

Nizovtsev and Novshek (2004) considered a two-period problem with two types of customers: high value and low value. Each type of customers includes some who has a portion that will not be satisfied by the product. They showed that extending the customer base in the second period by using a money-back guarantee can be optimal only if a monopolist faces an uncertain distribution of buyers. Within the second period, a money-back guarantee allows the monopolist retailer to discriminate between new and repeat customers, while a pure price reduction does not.

Chen and Bell (2012) examined how customer returns policies can be used to segment a firm's market into a dual-channel structure, by using a returnable channel and a non-returnable channel the retailer is able to enhance the firm's profit by segmenting customers using different returns policies.

In a duopoly model, McWilliams (2012) found that MBGs benefit the low-quality retailer while hurting the high-quality retailer. Although an MBG always benefits the monopolistic retailer, it is not necessarily beneficial to the retailer when there is competition involved. In a competitive environment in which a manufacturer supplies a product to a duopoly, Chen and Grewal (2013) examined how a new firm entering the market can use a full-refund policy or a no-refund policy as a competitive marketing strategy to compete against an existing firm that offers a full-refund policy. All the above studies use utility functions to derive demands.

With the assumption of aggregate demand, Chen and Bell (2009) discussed the impact of customer returns on a firm's pricing and ordering decision under a newsvendor framework. They assumed that the amount of returns is a proportion of the products sold and increases in proportion to the retail price. [These assumptions are supported by the empirical studies of Anderson et al. (2006) and Hess and Mayhew (1997).] Chen and Zhou (2014) investigated the loss-averse retailer's ordering policies for perishable product with customer returns. With the segmental loss utility function, the study developed the retailer's loss aversion decision bias and established the loss-averse retailer's ordering policy model. They found that both the risk-neutral and the loss-averse retailers' optimal order quantities depend on the inventory holding cost and the marginal shortage cost. Vlachos and Dekker (2003) extended the classical newsboy problem, which is to set the initial order quantity for a single-period product, to incorporate returns by assuming that customer returns were a fixed proportion of quantity sold.

Several studies have shown that a partial refund can discourage customers from returning purchased merchandise. Yan (2009) found that a partial return policy is the best choice for a retailer when the product is compatible with online marketing. Swinney (2011) considered the impact of returns policy on a firm's incentives to adopt a quick response strategy. The research identified conditions under which quick response increases/decreases profit (when a partial returns policy should be offered). Shulman et al. (2009) discussed how consumer purchase and return decisions are affected by a seller's pricing and restocking fee policy. Shulman et al. (2010) examined the impact of reverse channel structure on the equilibrium of partial return policy and profit of a bilateral monopoly. Shulman et al. (2011)

investigated the pricing and restocking fee decisions of two competing firms selling horizontally differentiated products. They assumed that consumers have heterogeneous taste for the products. They showed that restocking fees can be sustained in a competitive environment, but then stressed the importance of how restocking fees are also more severe when consumers are less informed about product fit and when consumers place a greater importance on how well products' attributes fit with their preferences.

A cluster of studies have also discussed supply chain coordination in the presence of a customer returns policy. Su (2009) studied the impact of full returns policies and partial returns policies on supply chain performance, with proposed strategies to coordinate the supply chain in the presence of consumer returns. Ruiz-Benitez and Muriel (2014) modelled and discussed supply chain coordination when the retailer is facing stochastic demand and an exogenously given retail price. Huang et al. (2014) investigated the design of incentive contracts to coordinate the supply chain in the presence of the secondary market as well as customer returns. Liu et al. (2014) discussed the issue of supply chain coordination when the refund amount is a decision variable. Their findings indicated that a buyback policy cannot achieve supply chain coordination. Since handling customer returns is part of reverse supply chain, some studies extended to examine the buyback contract between the product suppliers and retailers when the retailers face customer returns. Chen and Bell (2011) used a buyback policy with two buyback prices to coordinate a supply chain when the retailer faces both price-dependent demand uncertainty and customer returns. The retailer implements a full-refund policy. Su (2009) considered a supply chain consisting a manufacturer and a supply chain in which the retailer implements a partial returns policy. He showed that a buyback policy can enhance the supply chain efficiency. Yoo et al. (2015) discussed how the retailer should set price and customer returns policies when the Stackelberg manufacturer offers a buyback policy.

There have been few studies on customer returns with a sole focus on the fashion retailing industry. The research of Choi et al. (2013) is unique as it provides insight for mass customization retailers on making strategic decisions when deciding to offer no-refund or full-refund policies. Considering that mass customization (MC) fashion products usually cannot be returned, Choi et al. (2013) developed a model of stochastic fashion MC program with the consideration of consumer demand uncertainty. By modelling the optimization objective of a risk-averse MC fashion brand via a mean-variance approach, they identified the conditions under which a full-refund or a no-refund policy should be implemented. The numerical analysis in their research showed that whether the risk-averse MC fashion brand prefers to offer consumer returns with a full-refund policy or a no-return policy depends heavily on the demand-return correlation (DRC) parameter. Choi (2013) examined the optimal return service charge policy (a partial refund policy) by studying both risk-neutral and risk-averse MC companies. He revealed how the MC service provider's level of risk aversion affects the optimal return service charge policy. He also obtained the conditions under which it is optimal for some fashion companies to offer a zero-return service charge (free return with a full refund).

Hjort et al. (2013) empirically tested whether a "one-size-fits-all" strategy fits the fashion e-commerce business and evaluated whether consumer returns are a central aspect of the creation of profitability. They found that the segmentation of customers on the basis of both sales and return patterns can facilitate a differentiated service delivery approach.

3.4 Future Research Directions

As discussed above, fashion retailing is similar to other sectors of the retail industry in terms of customer returns-related issues, but it has some unique characteristics, including a high returns rate, short life cycle, fraudulent returns, and no allowance of returns for customized products. As pointed out in the literature review section, there are a very limited number of studies on customer returns management that focus on fashion retailing, but the implications and managerial insights from other returns-related studies can apply in fashion retailing. This section will identify some research gaps and opportunities which have been understudied in customer returns management, especially for fashion retailing.

In fashion retailing, future research directions could focus on the following: reducing amounts of customer returns, especially fraudulent returns; efficiently managing customer returns, including the attempt to salvage-returned products and provide customer returns policies that work in favour of fashion retailing.

There have been only very limited studies on customer behaviour in returning fashion products. The rapid advances in information technology and consumer analytics, however, allow retailers to collect and process large amounts of customer data (Acquisti and Varian 2005; Aydin and Ziya 2009) and more accurately gauge a customer's preference for products, returns behaviour, and willingness to pay (Wertenbroch and Skiera 2002). This provides opportunities for research on customer returns management. One research direction would be to find the correct size, colour, and style of fashion product for the customer, which could save the retailer and the customer from unnecessary trial and error through collect big data and business analytics techniques. As a result, the fashion retailing industry can significantly reduce customer returns. In addition, with the knowledge of customers' willingness to pay, the fashion retailer may be able to provide a menu that displays different prices with associated refund prices to customers. Development of an optimal menu could be a useful direction for future research.

Return fraud is a serious issue in the field of returns in the apparel industry. While there is no concise definition of return fraud, it can be said to include stealing merchandise from the sales floor and returning it, employees keeping receipts from previous sales and using them to process a refund later, copying and altering receipts, and merchandise borrowing (Kang and Johnson 2008). Return fraud is leading in many organizations to enact more restrictive returns policies. Kang and Johnson (2008) found that nearly 95 % of retailers were affected by consumers returning stolen merchandise for a refund, which costs retailers $16 billion and accounts for approximately 9 % of total returns a year. Evidence shows that consumers often exploit retailers' generous returns policies. In a study conducted by Rosenbaum and Kuntze (2005), nearly 20 % of consumers were found buying products with the specific intention of returning them after satisfaction was achieved, while nearly 18 % of all shoppers participate in some type of return fraud (Piron and Young 2000). Several US-based retailers, including Guess, Sports Authority, and Limited, have implemented shopper-tracking technology called "Verify" to detect abnormal return behaviours and allow retailers to reject returns that fail to meet certain criteria (Kang and Johnson 2008). Future research could consider methods and approaches for identifying fraudulent returns by studying customers' purchasing and returning behaviour.

Numerous studies have examined the existence of different types of returners: heavy returners, medium returners, light returners, and occasional returners (Foscht et al. 2013). Studies often focus on how to discourage fraudulent returns and fail to explore the consumer behaviour side. As such, further research could be conducted to examine how consumer satisfaction, dissatisfaction, and/or loyalty affect the quantity and frequency of product returns (Kim and Wansink 2012). Future research could also look into the consequences of return behaviour and whether it crosses product categories and purchase channels.

While there have been studies on how different degrees of return policy leniency affects various types of retailers, there is a lack in-depth analysis on how marketing mix variables affect retailers' policies. It would thus be interesting to study the interactive dynamics of retailers' return policies and other marketing mix variables (Kim and Wansink 2012) to isolate the effect of each marketing mix variable. For example, will price interact with return policy leniency to make certain products more likely to be chosen by consumers (Bahn and Boyd 2014)?

3.5 Conclusion

Product returns is an unavoidable aspect of any business that wishes to sell products to consumers and should thus be embraced by businesses to take advantage of the opportunities presented by product returns. With higher returns rates, serious fraudulent returns, and the short life cycle of fashion products, efficiently managing customer returns are crucial to the fashion retailing industry. With current technology and the hypercompetitive market, it becomes even more important to have outstanding returns policies that can satisfy consumers, because consumers' purchasing and returning decisions highly depends on return policies offered by retailers. Also, customers evaluate the product according to the advertised returns policy (Moorthy and Srinivasan 1995). Retailers can use their returns policies to better attract and retain consumers in an effort to develop long-term consumer loyalty. To do so, it is important for retailers to keep in mind the various factors that may influence consumers' evaluation of their returns policies and their effects on shopping behaviours. At the same time, it is important to minimize costs created by heavy returners or consumers that take advantage of lenient returns policies. Ultimately, retailers must walk a fine line between providing a satisfactory returns policy to consumers to attract and retain them, while ensuring that their returns policy does not negatively impact their profitability.

As discussed in this paper, there is a need for more research regarding customer return management, especially in the fashion industry, since academics and retail companies are beginning to understand that customer returns should no longer be looked at as a burden. Rather customer returns should be looked at for unexplored opportunities and improvements so that the topic can get a new positive status for customers, academics, and retailers.

References

Acquisti A, Varian HR (2005) Conditioning prices on purchase history. Mark Sci 24(3):367–381

Ahangarkolaei HE, Zandi MZ (2010) Using revenue management in multiproduct production-inventory systems: a survey study. Master's Thesis, Linköping University

Alptekinoglu A, Grasas A (2014) When to carry eccentric products? Optimal retail assortment under consumer returns. Prod Oper Manage 23(5):877–892

Anderson ET, Hansen K, Simister D, Wang LK (2006) How are demand and returns related? Theory and empirical evidence. Working paper, Kellogg School of Management

Autry CW (2005) Formalization of reverse logistics programs: a strategy for managing liberalized returns. Ind Mark Manage 34(7):749–757

Aydin G, Ziya S (2009) Technical note: Personalized dynamic pricing of limited inventories. Oper Res 57(6):1523–1531

Bahn KD, Boyd E (2014) Information and its impact on consumers' reactions to restrictive return policies. J Retail Consum Serv 21(4):423–425

Banjo S (2013) Rampant returns plague e-retailers. Wall Street J Dec 22

Barry C (2000) Happy returns: how to reduce customer returns-and their costs. Catalog Age, Nov 2000, 17, 12; ABI/INFORM Global, pp 108–110

Bernon M, Rossi S, Cullen J (2011) Retail reverse logistics: a call and grounding framework for research. Int J Phys Distrib Logistics Manage 41(5):484–510

Bhatnagar A, Misra S, Rao HR (2000) On risk, convenience, and internet shopping behaviour. Commun ACM 43(11):98–105

Biederman D (2005) Many happy returns. J Commer. Dec 5, 1–3 (New York)

Bonifield C, Cole C, Schultz RL (2010) Product returns on the internet: a case of mixed signals? J Bus Res 63(9–10):1058–1065

Boyaci T, Gallego G (2004) Supply chain coordination in a market with customer service competition. Prod Oper Manage 13(1):3–22

Cecilio D (2015) E-commerce is changing the fashion industry—it's time to catch up. The Guardian, Feb 6. http://www.theguardian.com/small-business-network/2015/feb/06/how-mobile-ecommerce-changing-fashion-industry

Che YK (1996) Customer return policies for experience goods. J Ind Econ 44(1):17–24

Chen J, Bell PC (2009) The impact of customer returns on pricing and order decisions. Eur J Oper Res 195(1):280–295

Chen J, Bell PC (2011) Coordinating a decentralized supply chain with customer returns and price-dependent stochastic demand using a buyback policy. Eur J Oper Res 212(2):293–300

Chen J, Bell PC (2012) Implementing market segmentation using full-refund and no-refund customer returns policies in a dual-channel supply chain structure. Int J Prod Econ 136(1):56–66

Chen J, Grewal R (2013) Competing in a supply chain via full-refund and no-refund customer returns policies. Int J Prod Econ 146(1):246–258
Chen X, Zhou Q (2014) Loss-averse retailer's optimal ordering policies for perishable products with customer returns. Math Probl Eng Article ID 831796, 5 p
Chen K, Kaya M, Özer Ö (2008) Dual sales channel management with service competition. Manuf Serv Oper Manage 10(4):654–675
Choi TM (2013) Optimal return service charging policy for a fashion mass customization program. Serv Sci 5(1):55–68
Choi TM, Liu N, Ren S, Hui CL (2013) No refund or full refund: when should a fashion brand offer full refund consumer return service for mass customization products? Math Probl Eng 23:1–15
Christopher M, Lowson R, Peck H (2004) Creating agile supply chain in the fashion industry. Int J Retail Distrib Manage 32(8):367–376
Chu W, Gerstner E, Hess J (1998) Managing dissatisfaction: how to decrease customer opportunism by partial refunds. J Serv Res 1(2):140–155
Cohen MA, Cull A, Lee HL, Willen D (2000) Saturn's supply-chain innovation: high value in after-sales service. Sloan Scl Manage 41(4):93–102
Creutz M, Larsson F (2012) Reverse logistics case study comparison between an electronic and a fashion organization. Master Thesis, Jonkoping University
Davis S, Gerstner E, Hagerty M (1995) Money back guarantees in retailing: matching products to consumer tastes. J Retail 71(1):7–22
Davis S, Hagerty M, Gerstner E (1998) Return policies and the optimal level of 'hassle'. J Econ Bus 50(5):445–460
Desmet P (2014) How retailer money-back guarantees influence consumer preferences for retailer versus national brands. J Bus Res 67(9):1971–1978
Ferguson M, Guide M Jr, Souza GC (2006) Supply chain coordination for false failure returns. Manuf Serv Oper Manage 8(4):376–393
Fernie J, Sparks L, McKinnon CA (2010) Retail logistics in the UK: past, present and future. Int J Retail Distrib Manage 38(11/12):894–914
Foscht T, Ernstreiter K, Maloles C III, Sinha I, Swoboda B (2013) Retaining or returning? Some insights for better understanding of return behaviour. Int J Retail Distrib Manage 41(2):113–134
Gecker R, Vigoroso MW (2006) Revisiting reverse logistics in the customer centric service chain. Benchmark Report. Aberdeen Group, Inc.
Grant BD, Lambert MD, Stock RJ, Ellram ML (2006) Fundamentals of logistics management, European edn. McGraw-Hill Education, New York
Grewal D, Munger JL, Iyer GR, Levy M (2003) The influence of internet-retailing factors on price expectations. Psychol Mark 20(6):477–493
Ha Y, Stoel L (2004) Internet apparel shopping behaviors: the influence of general innovativeness. Int J Retail Distrib Manage 32(8):377–385
Harris LC (2008) Fraudulent return proclivity: an empirical analysis. J Retail 84(4):461–476
Harris LC (2010) Fraudulent consumer returns: exploiting retailers' return policies. Eur J Mark 44 (6):730–747
Heiman A, McWilliams B, Zilberman D (2001) Demonstrations and money-back guarantees: market mechanisms to reduce uncertainty. J Bus Res 54(1):71–84
Hess J, Mayhew G (1997) Modeling merchandise returns in direct marketing. J Direct Mark 11 (2):20–35
Hess J, Chu W, Gerstner E (1996) Controlling product returns in direct marketing. Mark Lett 7 (4):307–317
Hewitt M (2008) Retailers: time to shout about your returns policy. Logistics Transp Focus 10 (8):58–61
Hjort K, Hellstrom D (2014) Experiment of consumer behaviour in fashion E-commerce—experience and insights. Lund University

Hjort K, Lantz B, Ericsson D, Gattorna J (2013) Customer segmentation based on buying and returning behavior. Int J Phys Distrib Logistics Manage 43(10):852–865
Huang X, Gu J, Ching W, Siu T (2014) Impact of secondary market on consumer return policies and supply chain coordination. Omega 45(3):57–70
Kang M, Johnson K (2008) Identifying characteristics of consumers who frequently return apparel. J Fashion Mark Manage 13(1):37–48
Kim S (2013) Online retailers tackle high rate of customer returns. ABC News, Dec 24, http://abcnews.go.com/Business/online-shopping-transactions-returned/story?id=21312312
Kim J, Wansink B (2012) How retailers' recommendation and return policies alter product evaluations. J Retail 88(4):528–541
King T (2004) Deshopping: retail perspective on the mismanagement and prevention of 'deshopping'. Eur Retail Dig 44:61
King T, Dennis C, McHendry J (2007) The management of reshipping and its effects on service: a mass market case study. Int J Retail Distrib Manage 35(9):720–733
King T, Dennis C, Wright LT (2008) Myopia, customer returns and the theory of planned behavior. J Mark Manage 24(1–2):185–203
Langley CJ, Coyle JJ, Gibson JB, Novack AR, Bardi JE (2008) Managing supply chains—a logistics approach, 8th edn. South Western College, Kentucky
Lee M (2009) An exploratory study of immoral consumer behavior: apparel retail salespeople's perspectives. Fashion Bus 13(3):62–76
Lei J, de Ruyter K, Wetzels M (2008) Consumer responses to vertical service line extensions. J Retail 84(3):268–280
Lele MM (1997) After-sales service: necessary evil or strategic opportunity? Managing Serv Qual 7(3):141–145
Li Y, Lei X, Li D (2013) Examining relationships between the return policy, product quality, and pricing strategy in on line direct selling. Int J Prod Econ 144(2):451–460
Liu J, Mantin B, Wang H (2014) Supply chain coordination with customer returns and refund-dependent demand. Int J Prod Econ 148(2):81–89
Martinez R (2009) Best practices in return management. http://multichannelmerchant.com/opsandfulfillment/best-practices-in-returns-management-01122009/
Masson R, Iosif L, MacKerron G, Fernie J (2007) Managing complexity in agile global fashion industry supply chains. Int J Logistics Manage 18(2):238–254
McCollough MA, Gremler DD (2004) A conceptual model and empirical examination of the effect of service guarantees on post-purchase consumption evaluations. Managing Serv Qual 14 (1):58–74
McWilliams B (2012) Money-back guarantees: helping the low-quality retailer. Manage Sci 58 (8):1521–1524
Melendez K (2013) How to reduce the cost of consumer returns posted. http://info.mkmdistribution.com/blog/bid/362624/How-to-Reduce-the-Cost-of-Customer-Returns
Millar M (2013) Return to sender: why the christmas boom can be a retail headache. BBC News, http://www.bbc.com/news/business-21148508
Min H, Ko HJ (2008) The dynamic design of a reverse logistics network from the perspective of third-party logistics service providers. Int J Prod Econ 113(1):176–192
Moorthy S, Srinivasan K (1995) Signaling quality with a money-back guarantee: the role of transaction costs. Mark Sci 14(4):442–466
Mostard J, Teunter R (2006) The newsboy problem with resalable returns: a single period model and case study. Eur J Oper Res 169(1):81–96
Murphy J, Benrs S (2015) Retailers found unprepared to meet "I want it now" consumer demand—New HRC advisory study uncovers key challenges within retail supply chain. Business wire—HRC advisory. Retrieved from: http://www.businesswire.com/news/home/20150915006463/en/Retailers-Unprepared-Meet-%E2%80%9CI-Now%E2%80%9D-Consumer-Demand#.VgP41pMsw-o
National Retail Federation Report (2014) Customer returns in the retail industry

Nike Returns (2015) Retrieved Nov 13, 2015, from http://help-en-us.nike.com/app/answers/detail/article/returns-policy/a_id/29785

Nizovtsev D, Novshek W (2004) Money-back guarantees and market experimentation. Int J Ind Organ 22(7):983–996

Norek CD (2002) Returns management. Making order out of chaos. Supply Chain Manage Rev 6 (3):34–42

Ofek E, Katona Z, Sarvary M (2011) 'Bricks & clicks': the impact of product returns on the strategies of multi-channel retailers. Mark Sci 30(1):42–60

Pei Z, Paswan A, Yan R (2014) E-tailer's return policy, consumer's perception of return policy fairness and purchase intention. J Retail Consum Serv 21(3):249–257

Petersen JA, Kumar V (2010) Can product returns make you money? MIT Sloan Manage Rev 51:85–89

Piron F, Young M (2000) Retail borrowing: insights and implications on returning used merchandise. Int J Retail Distrib Manage 28(1):27–36

Piron F, Young M (2001) Retail borrowing: Definition and retailing implications. J Retail Consum Serv 8(3):121–125

Pralle A, Stalk G Jr (2006) Returns: the ugly ducklings of retail. The Boston consulting group's report

Ramanathan R (2011) An empirical analysis on the influence of risk on relation- ships between handling of product returns and customer loyalty in E-commerce. Int J Prod Econ 130(2):255–261

Ratcliff C (2014) How fashion ecommerce retailers can reduce online returns. https://econsultancy.com/blog/65026-how-fashion-ecommerce-retailers-can-reduce-online-returns/

Richardson J (1996) Vertical integration and rapid response in fashion apparel. Organ Sci 7 (4):400–412

Rose A (1996) U.S. Patent No. 5, 930, 769. Washington, DC: U.S. Patent and Trademark Office. System and method for fashion shopping

Rosenbaum M, Kuntze R (2005) Looking good at the retailers' expense: investigating unethical retail disposition behaviour among compulsive shoppers. J Retail Consum Serv 12(3):217–225

Ruiz-Benitez R, Muriel A (2014) Consumer returns in a decentralized supply chain. Int J Prod Econ 147(1):573–592

Schmidt S, Kernan J (1985) The many meanings (and implications) of satisfaction guaranteed. J Retail 61(4):89–108

Schmidt R, Sturrock F (1999) Deshopping. The art of illicit consumption. Int J Retail Distrib Manage 27(8):209–301

Shieh S (1996) Price and money-back guarantees as signals of product quality. J Econ Manage Strategy 5(3):361–377

Shulman D, Coughlan AT, Savaskan RC (2009) Optimal restocking fees and information provision in an integrated demand-supply model of product returns. Manuf Serv Oper Manage 11(4):577–594

Shulman D, Coughlan AT, Savaskan RC (2010) Optimal reverse channel structure for consumer product returns. Mark Sci 29(6):1071–1085

Shulman D, Coughlan AT, Savaskan RC (2011) Managing consumer returns in a competitive environment. Manage Sci 57(2):347–362

Stalk G (2006) Customer returns top $10 billion in 2005: most Canadian retailers fail to capitalize on this key customer relationship. Canada NewsWire, Ottawa 1

Stuart J, Bonawi-tan W, Loehr S (2005) Reducing costs through improved returns processing. Int J Phys Distrib Logistics Manage 35(7):468–480

Su X (2009) Consumer returns policies and supply chain performance. Manuf Serv Oper Manage 11(4):595–612

Swinney R (2011) Selling to strategic consumers when product value is uncertain: The value of matching supply and demand. Manage Sci 57(10):1737–1751

The Economist (2013) Return to Santa—E-commerce firms have a hard core of costly, impossible-to-please customers. Retrieved Mar 1, 2014, fromhttp://www.economist.com/news/business/21591874-e-commerce-firms-have-hard-core-costly-impossible-please-customers-return-santa

Thompson E (2013) Unique characteristics of the fashion industry. http://inbloomautumnwinter2013.blogspot.ca/2012/05/unique-characteristics-of-fashion.html

Toktay LB (2003) Forecasting product returns, in business aspects of closed-loop supply chains. In: Guide VD Jr, Van Wassenhove LN (eds) Carnegie Mellon University Press. Pittsburgh, Pennsylvania

VandeVate J, Bedir G (2005) Many happy returns! Frontline solutions magazine, http://www.chromatographyonline.com/lcgc/Columns/Many-Happy Returns/ArticleStandard/Article/detail/153267

Vlachos D, Dekker R (2003) Return handling options and order quantities for single period products. Eur J Oper Res 151(1):38–52

Walsh G, Möhring M, Koot C (2014) Preventive product returns management systems: a review and model. In: 22nd European conference on information systems, Tel Aviv

Wertenbroch K, Skiera B (2002) Measuring consumers' willingness to pay at the point of purchase. J Mark Res 39(2):228–241

Wood SL (2001) Remote Purchase environments: the influence of return policy leniency on two-stage decision processes. J Mark Res 38(2):157–169

Xu L, Li Y, Govindan K, Xu X (2015) Consumer returns policies with endogenous deadline and supply chain coordination. Eur J Oper Res 242(1):88–99

Yan R (2009) Product categories, returns policy and pricing strategy for e-marketers. J Prod Brand Manage 18(6):452–460

Yoo SH, Kim D, Park MS (2015) Pricing and return policy under various supply contracts in a closed-loop supply chain. Int J Prod Res 53(1):106–206

Zabriskie NB (1973) Fraud by consumers. J Retail 48(4):22–27

Zappos.com customers enjoy free shipping, free returns, and 24/7 customer service! | Zappos.com (2015) Retrieved Nov 13, 2015, from http://www.zappos.com/shipping-and-returns

Part II
Theoretical Economic Models

Chapter 4
The Role of Quick Response in Accelerating Sales of Fashion Goods

Alejandro Lago, Victor Martínez-de-Albéniz, Philip Moscoso and Andreu Vall

Abstract Quick response has been proposed as an appropriate operational strategy to serve volatile markets. In fashion, postponing design, production, and distribution as much as possible may indeed reduce the uncertainty related to product success. In this paper, we provide an empirical study of the influence of lead time and sourcing origin on product success, based on data provided by a European fast fashion retailer. We provide a model of sales diffusion over time where product success is characterized by the *speed of sales*. We then evaluate how the speed of sales is influenced by the design time and the time-to-market of each particular product. We find that delaying the time of design is very beneficial, because it allows the firm to learn about fashion trends. The effect of time-to-market is more subtle. For a shorter time-to-market, speed of sales is considerably higher, but there is limited learning obtained by postponing design. In contrast, for longer time-to-market, speed of sales is lower, but the learning is higher, so for products designed late in the season, the speed of sales is similar to that of items with short time-to-market.

Keywords Dynamic demand model · Demand estimation · Lead time management

A. Lago (✉) · V. Martínez-de-Albéniz · P. Moscoso · A. Vall
IESE Business School, University of Navarra, Av. Pearson 21, 08034 Barcelona, Spain
e-mail: alago@iese.edu

V. Martínez-de-Albéniz
e-mail: valbeniz@iese.edu

P. Moscoso
e-mail: pmoscoso@iese.edu

A. Vall
e-mail: andreu.vall@jku.at

T.-M. Choi (ed.), *Analytical Modeling Research in Fashion Business*,
Springer Series in Fashion Business, DOI 10.1007/978-981-10-1014-9_4

4.1 Introduction

Reducing the lead time required to design, manufacture, and distribute the products is nowadays a powerful source of competitive advantage for fast fashion companies. Indeed, in apparel retailing, supply chain management has undergone a major shift of focus in the last decades: While historically supply efficiency was the primary concern, since the 1980s, the need for agility and speed to respond to demand trends has received increased attention both from industry and academia (Christopher et al. 2004).

Following introduction of pull production concepts in the early 1980s, which linked production decisions to evidence of downstream demand, a major milestone in supply chain management was quick response (QR), an apparel manufacturing initiative initiated in the USA in the 1980s (Hammond and Kelly 1990). Essentially, QR is a set of practices aiming at a better match between supply and demand. Through the years, QR practices have evolved and transferred to other industries and eventually have become a prevailing principle in supply chain management (Fisher 1997).

The primary objective of QR is to postpone ordering decisions as much as possible to benefit from better demand information and consequently reduce the risk of demand–supply mismatches. As such, QR allows a company (1) to make comparatively smaller investments in inventory and therefore mitigates the cost of stock clearances, as well as (2) reduces the opportunity costs of lost sales with to the possibility of faster replenishment. To achieve these goals, QR practices target the reduction of lead times and setup costs. Companies try to reduce lead times sufficiently so that production can be done in response to initial sales information (Fisher and Raman 1996).

Firms like Benetton first attempted to reduce lead time by postponing distribution through expedited logistics methods. Later on, QR practices were extended to postponement of production as well, typically achieved by nearshoring through local suppliers, and in recent years, of product design as well. This has enabled companies to respond even quicker to market trends and introduce new products very frequently, which is a key characteristic of fast fashion players such as Zara (Caro and Martínez-de-Albéniz 2015).

QR and fast fashion practices have been discussed extensively in the operations management literature. However, to the best of our knowledge, very limited empirical analysis exists on the practical impact on industry. There are unresolved fundamental questions such as whether a company that reduces its lead time sells more. The objective of this paper is thus to empirically establish whether lead time reduction is valuable, and if so, when and how much. We present results from an original data set, provided by a European fast fashion retailer for the fall–winter 2008 season. This single environment provides an experimental setting where product success is solely driven by variations in lead times and supply conditions. Specifically, the same type of product may be sourced from different types of suppliers, ranging from offshore low-cost suppliers to nearshore suppliers with significantly shorter lead times. This variability in the lead times for the same category of goods allows us to discriminate the value of QR.

To study the data, we develop a model of product success which is characterized by the *speed of sales*. This metric is defined as the percentage of total inventory available at the firm that is sold on average per unit time. It is a proxy for product success in our fast fashion context, for three main reasons. First, the faster a certain portion of a product is sold at full price, the less likely markdowns are needed at the end of the season because of excess inventory. Hence, a higher speed of sales is associated with a reduction in the excess-inventory cost present in traditional QR models. Furthermore, because new products are introduced continuously all along the season, including some that may resemble the current product, a higher speed of sales is associated with additional sales of similar items, similar to the reduction of lost-sales cost in traditional QR models. Second, retailers can introduce more frequently new products in their stores, so in a given period (e.g., a year), more units will be sold per square meter; hence, a higher speed of sales is linked to a lower opportunity cost of space per unit time that the end-of-season demand–supply mismatch approach in the traditional QR literature disregards. Finally, there is an indirect effect associated with higher speed of sales: The higher frequency of introductions will often lead to customer visiting the stores more often and typically buying more products (Bernstein and Martínez-de-Albéniz 2014). This is related to making strategic customers experience product scarcity (Cachon and Swinney 2011), which leads them to procrastinate less, i.e., postpone less often a purchase in the hope of later markdowns.

Our empirical analysis tests how this metric is influenced by the sourcing choices made on the different products, controlling for all other observable product features such as color, product type, average sales price, or cost. In particular, we focus on the role of two main variables: the time of design of the item, which can be delayed if an item is nearshored and will be important if fashion trends can be learned during the season; and the lead time or time-to-market itself, which becomes relevant if trends drift randomly, in which case a shorter time between design and sales should increase the success of the product.

We uncover several important findings. First, we find that the time of design strongly influences product success: Products designed later in the season tend to be more successful and sell faster. This confirms that fashion trends can be learned over the season and that products designed later will be more attractive to the customers. Second, we analyze the effect of time-to-market and find that short time-to-market makes products more successful, by penalizing items made offshore, which on average will sell slower. We also observe that the uncertainty in success rate is higher for offshore items, which suggests that there are more 'flops' in offshore products compared to nearshore ones. Third, we study the interaction between design time and time-to-market: We find that the learning obtained by delaying design is faster when time-to-market is higher, for offshore origins. In other words, for items with long time-to-market, being able to delay design time is very valuable, because the improvement of information about fashion trends can be translated into better products; in contrast, for items with short time-to-market, information cannot be improved by much when design time is delayed. As a result, speeds of sales of items designed early in the season are much higher for nearshore products compared to offshore ones, while they are indistinguishable when design happens later in the season. Hence, our work provides a first empirical evidence that QR, through delayed design and short time-to-market, can create significantly better products, even though they may be more expensive.

The remainder of this paper is organized as follows. Section 4.2 reviews the relevant literature. We present in Sect. 4.3 a sales diffusion model that we use to identify the parameters that determine product success. We then discuss in Sect. 4.4 the data and the empirical findings. Finally, Sect. 4.5 summarizes the conclusions, discusses some implications of this work, and lists some topics for further research.

4.2 Literature Review

Quick response (QR) is 'a set of practices designed to reduce the cost of mismatches between supply and demand' (Cachon and Terwiesch 2009). These practices include, for example, lead time reduction, postponement, product differentiation, and better forecasting. There is a broad literature on QR and fast fashion related to the present paper. On the one hand, starting from the mid-1990s, there are numerous papers presenting analytical models of QR practices and analyzing their implications. These papers suggest that the closer one gets to actual sales realization, the smaller the forecast error, and this is why QR leads to lower inventory costs and lower lost sales. We confirm this with our data, although we do not follow the traditional demand–supply mismatch approach. On the other hand, there is also some work of empirical nature focusing on the relationship between inventories and sales. We complement this stream of literature by providing an empirical estimate of the benefits of QR on the speed of sales.

4.2.1 Analytical Models

As companies aim at increasing their sales and margin through the introduction of new products in increasingly shorter cycles, they realize that these sales also become more unpredictable. Such innovative products are thus challenging in terms of inventory and/or stock-out costs, and companies will need for a responsive supply chain, rather than an efficient supply chain adapted to functional, stable, and lower-margin products (Fisher 1997). QR practices are the operational levers to enable such responsiveness. QR has been extensively studied and modeled in literature as a mean to reduce the risks of supply–demand mismatches.

Fisher and Raman (1996) and Iyer and Bergen (1997) proposed the first models that capture the improvement in the demand forecast between an initial ordering opportunity and a later one. The key trade-off is that in the later order, forecast is more reliable, but cost is higher. In centralized settings, QR combined with optimized decision making can significantly increase profits for a firm facing stochastic demand (Fisher et al. 2001). Fisher (2009) shows how forecasts can be improved by using initial sales data. More sophisticated demand models have also been proposed so that dynamic forecast updates can be incorporated, e.g., Heath and Jackson (1994). Song and Zipkin (2012) analyze such inventory planning problems with uncertain demand as a general case of the newsvendor problem. They observe that the dynamic programming formulation is closely related to a serial inventory system with stochastic demand over an infinite horizon. As a result, one can characterize the optimal policy and assess the value of additional demand information.

The postponement of supply to take advantage of better information can take different forms. First, one can consider the postponement of the distribution of products to stores, where the central pooling of inventory can provide higher sales with the same inventory levels. Caro and Gallien (2010) dynamically optimize shipments from a central warehouse to a network of stores and present how the proposed system was implemented at Zara. The implementation is further detailed in Caro et al. (2010). Second, it is possible to postpone manufacturing too. This practice is related to the decision on where the point of product differentiation should be located (Lee and Tang 1997). In particular, it may improve the service level and reduce inventories when dealing with product proliferation, as it will typically reduce manufacturing process complexity, and leverage better pooling effects. However, it may also require investments in the redesign of the products and processes (Feitzinger and Lee 1997). Third, product design is the last process that can be postponed. In other words, retailers can carry dynamic assortments to offer the most attractive product in each moment of the season. Caro and Gallien (2007) discuss how to optimally introduce new products based on sales expectations and demand learning opportunities and provide a closed-form policy that captures the key trade-offs between exploration and exploitation. Gaur and Honhon (2006) study product assortment and inventory decisions of a retailer under a locational choice model. They find that in some circumstances, static substitutions solutions can be a good proxy for the dynamic case problem and that in these cases, products of an optimal assortment should be equally spaced out at intervals to avoid substitution risks. Caro et al. (2014) and Çınar and Martínez-de-Albéniz (2013) study how to dynamically release products into the stores to keep assortments sufficiently attractive despite decaying interest from the part of the consumers.

In all the papers above, QR is shown to be useful in reducing the costs from demand–supply mismatches, through reduced inventory costs and lost sales. In contrast, we associate QR to higher speed of sales, which means that not only inventory costs are lower, but overall store sales are also higher (as new products can be introduced and sold), thereby reducing the opportunity cost of lost sales, and the costs of the store (rent, labor) will be better utilized. This effect is similar to the improvement in sales described in the dynamic assortment literature.

Finally, our dynamic model of sales borrows some elements from the work that links sales to inventory levels. Curhan (1973) relates availability of shelf space (which ultimately affects the amount of in-store inventory) to retail sales. Given that the amount of new products has grown faster that retail space, optimal space allocation has become a central issue for retailers. Corstjens and Doyle (1981) present an optimization model that takes into account elasticities to shelf space. Smith and Achabal (1998) and Caro et al. (2010) also provide retail models where sales are increasing concave functions of inventory levels. Balakrishnan et al. (2004) and Baron et al. (2011) study how to manage inventory in such setting. Finally, supply chain coordination issues have also been studied in this setting. Wang and Gerchak (2001) show that in order to coordinate a manufacturer and a retailer, the manufacturer should not just push the wholesale price lever, but offer as well an inventory holding cost subsidy to the retailer. Martínez-de-Albéniz and Roels (2011) study the competitive dynamics that result from such a shelf scarcity and demonstrate that, with wholesale price contracts, the inefficiencies from decentralized decision making are limited and can be fully solved through pay-to-stay fee contracts.

4.2.2 Empirical Models

In contrast with all the analytical papers cited so far, to the best of our knowledge, there is scarce empirical work on QR practices in fashion retailing besides estimates on the value of QR based on data in Fisher and Raman (1996), and business school cases (e.g., Ghemawat and Nueno 2003; Ferdows et al. 2002; McAfee et al. 2004; Caro 2011).

Most papers related to our work examine the effect of inventory (lower under QR) on sales. Chen et al. (2005) examine across different industries the relationship between inventory management practices and financial performance during 1981–2000. They find that the companies that have very high inventories show poor stock returns in the long term but that this relationship does not hold in the opposite case, as the firms with the lowest inventories show average stock returns. Gaur et al. (2005) empirically study inventory turnover at US retailers between 1987 and 2000. As turnover varies widely across retailers and over time, they conclude that the metric can be improved by adjusting for changes in gross margins, capital intensity, and sales surprise (which they define as the ratio of actual to expected sales for the year). These parameters condition significantly effective inventory productivity of a firm. Similarly, Kesavan et al. (2010) conclude that sales forecast of US retailers can be improved by incorporating exogenous variables into the estimation, such as inventory and gross margin. They test their model with historical data of inventory and gross margin and show that the model generates more accurate sales forecast than the forecasts of equity analysts.

The drivers for inventory have also been studied. Rumyantsev and Netessine (2007) use aggregate inventory data to measure the relationship between variables as demand uncertainty, lead times, gross margins, and firm size on the level of inventory. They find evidence that companies facing a higher demand uncertainty, longer lead times, and higher margins carry larger inventories. They also confirm that companies leverage economies of scale in inventories, as larger firms have relatively less inventory. Olivares and Cachon (2009) test empirically the key assumption that an increase in sales will lead to a less than proportional increase in inventory, with car dealer data. However, their results indicate that dealers carry more inventory when they face stronger competition, because a higher service level will compensate the potential sales reduction due to more competition. This is in contrast with Amihud and Medenelson (1989), where public data on manufacturing firms is used to estimate the effect of market power (which they proxy by margins and market shares) on inventory levels and variability. It is found that when market power decreases (i.e., competition intensifies), firms tend to lower their inventories.

Finally, the relationship to consumer choice has also been explored. Musalem et al. (2010) present a structural demand model that can capture the effect of stock-outs on customer choice, which they test empirically for consumer products. The model illustrates how the amount of lost sales induced by stock-out can be quantified. Ton and Raman (2010) analyze store data to study the effects of product variety and inventory levels on store sales in electronics retailing. They show that an increase of product variety and inventories leads to more misplaced products, which again decreases store sales.

In our paper, we take a more direct approach: We consider an alternative metric of product success and the speed of sales, and then study how it is driven by lead time choices.

4.3 The Model

4.3.1 A Continuous Inventory-Dependent Sales Model

Consider a fast fashion retailer that sells fashion goods over a network of stores. To sell a given product i, the firm follows a process that starts with the design of the product, continues with a sourcing decision where a manufacturer (in-house or external) is selected together with a production quantity Q_i and ends with the reception of the merchandise at a central warehouse from which it is distributed to the stores. After these steps, the product is displayed in the stores, seen by the customers, and purchased if they find it good enough.

We focus on fashion products, where there is significant uncertainty on the success of the product. Traditionally, due to the relative short life cycle of these goods compared to the design-to-store lead time, the literature has used the newsvendor model to capture such risks. Namely, a stochastic demand is realized; either excess inventory is salvaged or inventory is sold out; and some sales are lost. We claim that this model is inadequate in fast fashion as it overlooks some important elements. First, this is intrinsically a multi-period model, and sales do occur over many weeks. In fact, generally it is possible to sell all the inventory of a given product if the retailer keeps it in the store long enough. Second, demand is not the same when the product is introduced or when the product has been in the store long enough, because, among other things, inventory availability is not the same, which influences how the product is displayed and how much consumer attention it captures (Caro et al. 2014; Çınar and Martínez-de-Albéniz 2013). Third, retailers do not keep all the products in the store for the same amount of time, which implies that two products that have the same amount of sales may be very different if one has only stayed in the store for a few weeks while the other has been there for months. As a result, we build a more detailed sales model that integrates these facts. Our data will then be used to calibrate the model, and we will be able to test some hypotheses on what makes a product successful.

For product i, we let $t = 0$ be the time of introduction of the product in the stores. In every time interval $[t, t + dt]$ after the product introduction, sales are materialized. We hence consider a continuous-time sales model, as in Balakrishnan et al. (2004), which provides tractability for the analysis and can then be easily extended to discrete sales periods (weeks in our data). Furthermore, this is a realistic choice when the lot sizes are sufficiently large, as is the case in our data (hundreds to thousands of units per product).

The amount of infinitesimal sales is a function of the inventory available in each of the stores, together with a random component that essentially depends on how much the visitors of the store like the product's design, quality, and price. Interestingly, the random element in a product's success turns out to be highly correlated across time. We observe this phenomenon in our data, and most of the existing literature suggests a similar behavior. For instance, Caro and Gallien (2007) assume that a product has a certain 'true' sales rate common to the store network that needs to be learned over time; Fisher (2009) observes that early sales are highly predictive of future sales. This implies that, even though sales may be unpredictable before the introduction of a product, once the product is introduced into the stores and some early demand is revealed, most of the demand uncertainty is resolved.

Hence, we include a random element $\tilde{\epsilon}_i$ that is unknown at the product design-sourcing stage, but is revealed by the time the product is in the store. For simplicity, we assume that sales become deterministic for $t > 0$, although we show below how to include a random component after the product is in the store.

Furthermore, we see in our data that the sales of any given product are also highly correlated across stores. In other words, even though the success of skirts may be higher in France than in Austria, a given skirt will have either relatively high sales in France and Austria or relatively low sales in France and Austria. This allows us to consider network sales s_{it} rather than focusing on individual store sales. Specifically, we use the following modeling of the infinitesimal sales quantity

$$s_{it} = \alpha_i q_{it}^{\beta_i} \tag{4.1}$$

where q_{it} is the network inventory level and $\alpha_i > 0$, $0 \leq \beta_i \leq 1$ product-dependent parameters that may depend on $\tilde{\epsilon}_i$. This formulation that links sales to inventory is common in the literature, e.g., Corstjens and Doyle (1981), Wang and Gerchak (2001), Balakrishnan et al. (2004), or Martínez-de-Albéniz and

Roels (2011). It captures the fact that better displays through higher inventory quantities lead customers to pay more attention to the product and hence increase the purchase probability. It also suggests that there are decreasing returns from carrying more inventory: Doubling inventory only increases expected sales by a factor $2^{\beta_i} \leq 2$. Although the functional model is different, the same characteristics are present in the piecewise-linear relationship between sales and inventory of Smith and Achabal (1998).

Furthermore, it is worth pointing out that this formulation is robust to aggregation, i.e., valid both at the store and the network level. Indeed, if the sales-inventory sensitivity β_i is common to all the stores and the retailer distributes inventory in stores so as to maximize total sales over the network, then the sales over all stores k are as follows:

$$\max_{q_{itk} \geq 0 | \sum_k q_{itk} = q_{it}} \sum_k \alpha_{ik} q_{itk}^{\beta} = \alpha_i q_{it}^{\beta}$$

with $\alpha_i = \left(\sum_k \alpha_{ik}^{1/(1-\beta)}\right)^{1-\beta}$. This means that the retailer can distribute the inventory appropriately such that the marginal value of placing an additional item in a store is the same across all the stores. Then, the relationship (1) between inventory and sales exists both at the store level and the network level, which makes our focus on the network robust. For simplicity, we focus on the case $\beta_i = 1$ in this paper.

Now that we have defined sales as a function of inventory, we can use the inventory balance equation

$$\mathrm{d}q_{it} = -s_{it}\mathrm{d}t \tag{4.2}$$

to establish that

$$q_{it} = Q_i e^{-\alpha_i t}. \tag{4.3}$$

Note that the analysis can be easily extended to include an uncertain element in the sales trajectory: $\mathrm{d}q_{it} = -q_{it}(\alpha_i \mathrm{d}t + \sigma_i \mathrm{d}W_t)$, where W_t is a Wiener process. This would allow us to use our model for random demand processes too.

We shall refer to α_i as the *speed of sales* for product i. This metric provides a relative-to-inventory measure of how fast sales are materializing. This is appropriate because in our fast fashion context the key constraint is store space, that is, consumed proportionally to the amount of inventory displayed, as pointed out in the introduction. As a result, the speed of sales α_i indirectly measures how much each square meter of store can turn per unit time when carrying product i. Since for fast fashion retailers new products are being introduced continuously, it is always preferable to carry products with higher speeds of sales, so sales and margins are maximized, while inventory and store costs remain constant and discounts are minimized.

4.3.2 Relationship Between Sales Parameters and Operational Choices

Thus far, we have characterized the inventory trajectory after a product is introduced, which is exponentially decreasing over time. We turn here to the question of how the speed of sales for product i, α_i, is formed. Recall that it depends on some observable product $FEATURES_i$ but also on a random component $\tilde{\epsilon}_i$ that describes how much consumers like the product, as determined by some unobservable design characteristics, which is independent of observable product characteristics. Specifically, the variables $FEATURES_i$ include product category, color, average sales price, and cost.

In addition, we are particularly interested in understanding how the retailer's operational choices affect the realization of the speed of sales parameter, specifically the choice of suppliers and sourcing location. This drives the two variables on which we focus: the effective design-to-store *time-to-market*, T_i; and the design time D_i, defined as the first time the product was introduced in the store minus the time-to-market.

In other words, we pose the following relationship:

$$\alpha_i = f(D_i, T_i, FEATURES_i, \tilde{\epsilon}_i). \tag{4.4}$$

Since we do not know the time-to-market T_i exactly, we shall use the region of origin as a proxy for it. We shall consider the following regions (the average T_i for each region is provided): North Africa (4 weeks), East Asia (16 weeks), South Asia (12 weeks), East Europe (4 weeks), and West Europe (2 weeks). Furthermore, we shall also assume that the random element $\tilde{\epsilon}_i$ is exogenously generated and that it is uncorrelated with T_i, D_i. This is reasonable, since as we explain later in Sect. 4.4, we do not expect that sourcing areas are self-selected based on an a priori evaluation of the product sales forecasts.

In a fashion context, we expect that a shorter time-to-market allows the company to react faster to revealed consumer tastes. As discussed earlier, fashion is characterized by a high degree of uncertainty and volatility of consumer preferences. A long time-to-market forces companies to commit to design choices well before any real consumer preferences about colors, styles, or specific designs are revealed. On the other hand, a shorter time-to-market may enable, once some consumer preferences about certain product characteristics are known, to design and manufacture the product according to these preferences quickly enough to bring it to market before changes in preferences occur. Hence, a shorter time-to-market would result on a reduction on the uncertainties with customer tastes and will then lead to more successful products: In other words, products that are procured from closer locations and have a shorter T_i and a later D_i, should have a higher (i.e., faster) speed of sales α_i. This argument can be elaborated further. It is generally agreed in the literature that fashion trends can be only roughly guessed at the beginning of the season by a company, and they are only gradually revealed as products are being sold and the season advances. One could imagine two extreme ways in which these season trends evolve and are revealed.

In one extreme, trends are relatively stable and converge during the season (e.g., gooseneck shirts with Chinese motives start to be sold and as they are seen they become more successful and by the end of the season most people wear them), as in Eppen and Iyer (1997) or Fisher and Raman (1996). In this case, the company will gradually observe the trend as the season advances. Products designed later in the season will take advantage of this learning and hence will likely be more successful on average. In this case, a later design time will be correlated with a higher speed of sales. Of course, products designed later in the season will increasingly be produced in nearshore origins so that they can arrive on time to the stores. Hence, one would expect that products from nearshore origins are more successful, but this is exclusively due to the fact that they are designed later, i.e., with higher D_i.

On the other extreme, however, one could imagine that season trends are short-lived and that they are continuously evolving on a unpredictable manner during the season length (i.e., Chinese motives are being fancy this week, but next month Madonna will promote Scottish prints that substitute Chinese motives), as in Caldentey and Caro (2010). In this case, when some trend is spotted, one would like to design, produce, and bring the product to market as soon as possible in nearshore origins. However, there will be no convergent learning about the season trends. In this case, one can imagine preferences as moving in a random walk manner with no clear convergence. Products such that the lead time between design and introduction is short are able to take advantage of the trend before it dies and hence will be more successful. But a later design time will need not be influencing the success per se, as trends evolve during the season. Hence, in this case, products from nearshore origins would be more successful because of their shorter lead time, independently of the time of design or introduction, i.e., with lower T_i.

According to industry experts, reality may well be a combination of these depending on the season and product categories. Hence, our model considers both the time of design to capture season learning, and the origin location (as a proxy for design lead time) to capture design adaptation to short-lived trends. Being this the first empirical paper of this type, our scope will be restricted to show that nearshoring has a positive effect via a possible combination of each of these effects. This leads to the formulation of several empirical questions that we can test through the data. One may additionally try to estimate the relative importance of one of the effects against the other, but unfortunately our data do not allow to do that, because T_i and D_i exhibit high correlation.

4.3.3 Hypotheses

We are especially interested in evaluating the impact of the D_i and T_i on the speed of sales α_i. This leads to three hypotheses.

H1. *A later time of design $D_i$has a positive impact on speed of sales α_i, across all origins.*

According to H1, we expect to find that items designed later in the season sell faster, because the retailer can learn about customer tastes and thereby design more attractive products.

H2. *A shorter time-to-market* T_i, *given the same time of design* D_i, *has a positive impact on speed of sales* α_i.

According to H2, we expect to find that items with shorter time-to-market tend to be more successful. The proximity to market allows the firm to follow the market trends more accurately and therefore to offer products that sell faster. In contrast, a longer time-to-market means that there will be more time between design and distribution for customers to change their preferences. Hence, H2 would confirm that there is some uncertainty associated with a trend 'drift,' independent of the time of design and thus not associated with in-season learning of market trends (which is the focus of H1).

Furthermore, we shall consider one additional hypothesis on the interaction between D_i and T_i.

H3. *The marginal increase of speed of sales* α_i *due to a shorter time of design* T_i *is reduced when the time of design* D_i *increases.*

H3 means that the advantage of shorter time-to-markets is reduced as the season progresses. In other words, products designed in long lead time origins benefit more from the learning accumulated during the season; hence, for products introduced at the end of the season, the advantage of a short time-to-market is not as important.

4.4 Empirical Analysis

To validate the model presented in Sect. 4.3, we use data from a fast fashion retailer. We describe next the practices of the company and the details of the data set, and then present the empirical analysis to test our hypotheses.

4.4.1 Industry Setting

We use data from a fast fashion retailer based in Spain that wished to remain anonymous. We will call the firm F. The company's business model is similar to that of the Inditex group (holding of many brands, including Zara), the company that coined the term *fast fashion*, and is described in Caro and Martínez-de-Albéniz (2013, 2015). Fast fashion retailers have not only become market leaders (Keeley and Clark 2008), but have also dramatically changed the competitive dynamics in the industry. These companies are able to offer very fashionable products at affordable prices. This requires carrying a very dynamic store assortment that is able to adjust the products in the store to emerging demand trends. Such strategy allows fast fashion retailers to differentiate from traditional fashion retailers that introduce a new collection every six months only but also from newly emerged low-cost apparel retailers as Uniqlo or Kiabi, which focus their offering on basic, less fashionable items (e.g., plain-color T-shirts). The fast fashion practices have been described extensively in the business press, for example in Foroohar (2006), El País (2008, 2011), Butler (2013), or Berfield and Baigorri (2013). In addition, many case studies have been written, in particular on Zara, considered the current best practice (e.g., Ferdows et al. 2002; Ghemawat and Nueno 2003; McAfee et al. 2004; Caro 2011).

Fast fashion retailers have made supply chain management a centerpiece of their business. They have developed a much tighter control of their supply chains, structurally aiming for a vertical integration of most of the sourcing, production, distribution, and retailing tasks, and even assuming economic ownership of large portions of the value chain (e.g., warehouses and stores). The priority is very often put on execution speed (i.e., minimizing the time from design to store), rather than on cost. For instance, when choosing sourcing locations, nearshoring may be preferred. Furthermore, to manage their supply chains, they have typically implemented end-to-end information systems that provide real-time visibility of sales at the item level, which can be very useful for distribution, sourcing, and design decisions. This does not mean that these systems are complex: Zara is known for comparatively basic IT solutions where business needs are always put first before technological possibilities (McAfee et al. 2004).

It is important to note that, while the fast fashion model makes these companies different, it is not always the most appropriate strategy for any type of product being sold (Fisher 1997). Specifically, it is applied almost exclusively for fashion items, i.e., those with a design with high fashion content, that is, truly new compared to past designs. For more basic items where no real novelty is present in the design, a traditional, efficient supply chain is used, even at Inditex. As a result, to cope with the distinctive requirements of basic and fashion items, two distinct supply chain models are operated in parallel: A very efficient supply chain with a primary focus on cost optimization delivers their basic items that have comparatively constant demand patterns and high volumes; and an agile supply chain produces the trendier products, with the priority being speed and responsiveness to demand.

F also operates a similar dual-supply chain structure. This setting is ideal for the study of quick response, because it provides a controlled experiment where different products are produced with a long or a short time-to-market. Of course, the products may have different characteristics, but these will be controlled for in the empirical analysis. To be able to extract insights, we nevertheless require a certain amount of variability: to discriminate between long and short time-to-market, we need that products with the same characteristics (category, price, features, etc.) are supplied by both supply chains. In other words, we need that F sources in Asia (long time-to-market) and Europe or North Africa (short time-to-market) the same type of products. Fortunately, this is so for most of the product categories. However, for certain categories, only one origin is used, due to country specializations (e.g., all belts at F are made in China), which means that we cannot discriminate the value of QR from the data, and as a result, we do not include these in our analysis.

4.4.2 Company Background

F started operations in the year 2000 as a multi-brand store. While initially, F had decided to offer around 50 % of high-street brands, 25 % of low-cost own brands, and another 25 % of newly developed fashion brands, F decided in 2005 to focus on its own brands completely. These brands, grouped in 5 key brand families, are controlled by F from the design to the store. Our data are from the fall–winter 2008 (FW08) season, from April 2008 to January 2009, a season representative of the new strategy of the firm. We describe F's situation in 2008 next.

As of 2008/2009, the retail footprint of F is made of 141 stores with both owned and franchised stores, which are treated identically (same distribution and retail processes). Furthermore, the company complements its store network through corners, wholesalers, and shop-in-shops. Owned stores represent more than half of total sales, franchises slightly more than a third, with rest being corners, wholesalers, and shops-in-shops. Average revenue per point of sale and year is about €200,000, although owned stores and franchises may have significantly higher sales that the other points of sales. F has retail presence in 7 countries: The majority of sales originate in Spain, but it also has stores in France, Portugal, Italy, Greece, Romania, and Russia. Target customers are young people, primarily women, 15–30 years old, who are value-oriented but looking for highly fashionable products.

The company sells a wide range of fashion products. During a typical season (spring–summer and fall–winter), including FW08, F offers between 500 and 1000 different stock-keeping units (SKUs) , defined as a unique combination of model and color. The products are grouped in product families, such as T-shirts, dresses, bags, and shirts. T-shirts are the main product categories representing around 15 % of sales and 25 % of the SKUs. The remaining revenues and SKUs are relatively equally distributed among the other product families, with an average product family representing around 6 % of SKUs and 8 % of sales. The average price of a product is in the range of 10–20, and the company sells in the range of 1 million units per year in total.

The products are sourced worldwide from 14 different countries, such as Bangladesh or China, but also some production is allocated to quick-response suppliers in Spain, Portugal, Bulgaria, or Morocco. Production quantities are determined in one shot, after a design has been completed, and no additional production of the same product is ever made. However, new products are developed in season that can be similar to successful existing products, but they are treated as different SKUs. The firm in fact learns about customer preferences during the season: Designers permanently track fashion trends and customer preferences. There is a single, centralized design department that is organized around the main product families. Thus, the firm may be able to make these newly designed products more attractive, which should result in a higher speed of sales. Different types of products are sourced from different types of suppliers, ranging from offshore suppliers to nearshore suppliers with significantly shorter lead times. The cost of an item, which is a control variable in our study, may of course depend on the location

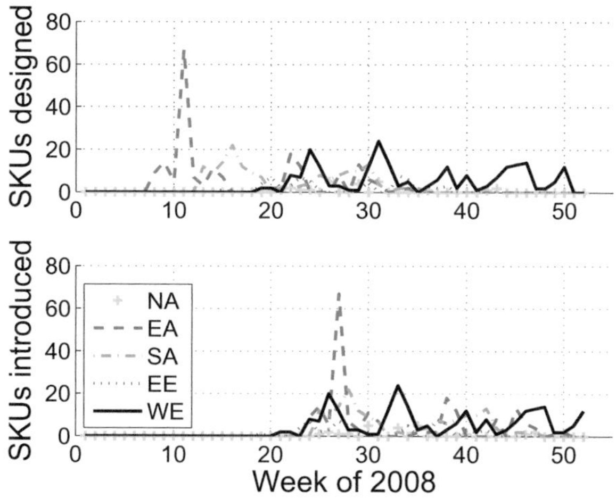

Fig. 4.1 Number of SKUs designed (*top*) and introduced (*bottom*) by week and region

of the supplier, but nearshore goods are not systematically more expensive: For instance, the average cost of a shirt or a pair of trousers is higher in Europe than in Asia, although the opposite is true for dresses. In both cases, the differences are minimal.

The process of deciding which supplier makes which product works as follows. The firm uses a master calendar for their product category introductions over the season, with the aim of having a balanced assortment in the stores. In the calendar, the decision of using an offshore versus nearshore supplier is specified, so as to be able to plan design, production, and distribution activities (including early communication of capacity requirements to suppliers, before products have been designed). These introductions are depicted in Fig. 4.1. It is worth highlighting that the amount of products being introduced at the beginning versus in the middle of the season has remained more or less stable over several seasons. As a result, when F decides to introduce a product at a certain date, the choice of offshore versus nearshore has already been fixed upfront. Thus, it should not create a self-selection problem in our analysis, although we discuss the impact of potential endogeneity in Sect. 4.4.7.

Hence, when the design for a specific product starts, its lead time requirement is already decided. The moment at which this occurs is what we call design time. It takes place with the latest available information about trends. After the main design phase is complete, procurement managers are able to quickly make initial cost estimates of the new products and track available capacity at the suppliers. Once the managers select a prototype for production, designers make final adjustments in colors and materials and coordinate specifications with the chosen supplier, that is, only eligible if it can meet the required lead time. The order is then placed. Finally, after the prespecified lead time, the items are introduced in the stores.

4.4.3 Data Set Description

We use a company database with information at the SKU level. Every SKU is uniquely defined by the model (design) and the color. That implies that different sizes are not considered different SKUs. Since, the distribution of sizes is fairly homogenous within each product, no bias is introduced with this aggregation. Data were properly filtered to remove inconsistencies. Only the ten product families with higher percentage of revenue over the total annual revenue and with at least three different origin areas are included, namely *scarf, shirt, T-shirt, knitwear jacket, cardigan sweater, jersey, trousers, shorts, top,* and *dress*. Furthermore, the products with very small purchase volumes (Q less than 5 units) were removed. These products are not representative of the company usual batch size (see Table 4.1 for relevant statistics of the products). Similarly, a few items had total sales higher than the purchase quantity, which we removed as well. After these filters, the final number of SKUs is 653.

The database includes on the one hand, sales volumes in units, per SKU, day and point of sales; on the other hand, the features of each SKU include family, color, average sales price per week, and purchasing cost per unit as well as the total purchased volume in units (inbound quantity from supplier into F), and the day of arrival of the product at F's unique warehouse. As we mention above, we take this day of arrival to the warehouse as the time of introduction of the product since cross-docking operations make the available the product at the store in the next few days.

Note that family *t-shirt* shows the highest variety of SKUs, while the families' *shorts* and *top* show little variety. For all the families, the average batch size is far

Table 4.1 Descriptive statistics of the key variables by product family

Product family	Number of SKUs	% overall SKUs	Average (std. dev.) of batch size	% of SKUs sourced in				
				EA	SA	NA	EE	WE
Cardigan sweater	40	6	690 (242)	50	15	0	0	35
Dress	84	13	701 (290)	31	12	2	6	49
Jersey	83	13	752 (402)	30	47	0	0	23
Knitwear jacket	86	13	900 (694)	65	12	0	0	23
Scarf	41	6	286 (258)	76	17	0	0	7
Shirt	45	7	841 (168)	4	7	9	58	22
Shorts	7	1	738 (138)	0	43	14	29	14
Top	12	2	615 (438)	33	0	0	33	33
Trousers	60	9	1223 (1038)	2	20	27	18	33
T-shirt	195	30	903 (726)	14	36	0	5	45
All categories	653	100	824 (633)	30	25	3	9	33

from the threshold of 5 units below which the product was eliminated. Generally, there is variation in the batch size, as shown by the standard deviations, although there was no particular association between batch size and other indicators such as product family or origin. Furthermore, each SKU has a unique sourcing origin. Countries of origin are classified into regions that share similar practices and delivery lead times: East Asia (EA: China, Vietnam), South Asia (SA: Pakistan, India, and Bangladesh), North Africa (NA: Morocco), East Europe (EE: Bulgaria, Romania, Turkey), and West Europe (WE: Spain, Portugal, France, and Italy). Since exact lead times are not available at the SKU level, we use these regions as proxies for supplier lead times and hence time-to-market. As stated in Sect. 4.3.2, the representative lead times for each region are 2 weeks for WE, 4 weeks for EE and NA, 12 weeks for SA, and 16 weeks for EA, respectively. As shown in Table 4.1 for each product category, nearshore and offshore origins are represented by a minimum amount of SKUs (at least 20 %), with the exception of *scarfs*.

The previous Figure 4.1 depicts the number of SKU designed and introduced as the season progresses, per week. The first introduction for the fall–winter season takes place in April, on week 17 (earlier than one may think), which means our timeline spans a total of 38 weeks. One can observe that product introductions tend to be lumped around some specific weeks for each specific origin (as shown by the spikes in the figure). Nevertheless, each origin has products introduced regularly through most of the season. In addition, one can observe that there is a strong collinearity between design time and time-to-market: For items which have late design times, the time-to-market must be shortened to launch them in timely manner. Furthermore, the retailer also tends to introduce earlier in the season a larger share of items made in Asia, with larger time-to-market; hence, products from these origins tend to be designed earlier as well. These observations imply that when incorporating both D_i and T_i in our regressions, we will have to work with cross-effects.

Unfortunately, for regions North Africa and East Europe, the number of new SKUs in any week is small. This does affect some of our empirical conclusions regarding these regions. Namely, when analyzing the effects of both design time and time-to-market on sales, we will need to consider aggregating our regions into broader categories to avoid spurious effects: nearshore (North Africa, West Europe, East Europe) and offshore origins (East Asia and South Asia). With this broader aggregation, we will have enough products which are introduced from offshore origins in weeks 20–30, which have been designed and produced starting between weeks 4 and 18, that is, when seasonal customers' preferences have already been revealed at the beginning of the season. Thus, this allows us to test whether there is a learning effect even for offshore origins. Hence, despite the loss of granularity, this approach will allow us to make robust inferences about the impact of time-to-markets (defined as either nearshore or offshore procurement) on sales as the season progresses and learning occurs.

4.4.4 Computation of α_i

In a fast fashion context, few measures of sales performance have been developed or tested. In our data, total production quantities are small, and there is no product replenishment. Hence, absolute measures such as total season sales volume are not adequate to measure performance since sales volumes are censored due to small quantities produced. Sell-through (i.e., the ratio of sales to purchase volumes) is also problematic because it is difficult to define an adequate selling period where the sell-through can be observed. Indeed, a large number of products tend to sell out eventually and replaced by new similar models, while, for others selling poorly, prices may be reduced at some point to accelerate sales. As a consequence, depending on the time frame used, products may eventually show different behavior. To avoid these problems, we shall consider as a measure of sales performance the speed of sales α_i, described in Sect. 4.3. We will nevertheless explore sell-through directly in Sect. 4.4.6.

To compute the value of α_i for product i, we first define the sell-through as the ratio

$$ST_{it} = \frac{SALES_{i[0,t]}}{Q_i} \tag{4.5}$$

where t is the time, measured in weeks, elapsed since the initial date when the product arrived to the warehouse, $SALES_{i[0,t]}$, is the cumulative sales of product i up to time t in all stores, and Q_i the total inventory purchased of product i. Of course, $0 \leq \mathrm{ST}_{it} \leq 1$.

From Eq. (4.3), $\ln(1 - \mathrm{ST}_{it}) = -\alpha_i t$. Furthermore, since by construction $\mathrm{ST}_{i0} = 0$ and hence $\ln(1 - \mathrm{ST}_{i0}) = 0$, we define α_i as

$$\alpha_i = -\frac{\sum_{t=1}^{W} t\ln(1 - \mathrm{ST}_{it})}{\sum_{t=1}^{W} t^2}. \tag{4.6}$$

We choose a window W to be the minimum of 17 weeks (4 months) and the time where the discount season starts, i.e., at the end of December (week 52 is the last week to be included). The reason we limit the number of weeks is that, after a long time, inventory starts running low and so is the sales quantity, and it may add noise to our estimator. Instead, we focus on the initial period when the product is introduced. Note however that our results are very robust to all other time windows that were tried. Furthermore, we limit our attention to the weeks where items are typically sold at full price, see detailed explanation below.

The definition of α_i in Eq. (4.6) deserves some explanation. One can see that it is a weighted average of $\ln(1 - \mathrm{ST}_{it})/t$, i.e., the rate at which sell-through decreases. The weights are proportional to t^2, which means that data from the later weeks are more important in determining α_i, although at these times $\ln(1 - \mathrm{ST}_{it})/t$ is also much smaller. The formula in (4.6) is in fact the parameter that best fits an exponential sell-through evolution over time. Specifically, α_i minimizes the squared deviations between $\ln(1 - \mathrm{ST}_{it})$ and $-\alpha_i t$, as in the ordinary least squares approach:

$$\min_{\hat{\alpha}_i \geq 0} \sum_{t=1}^{W} (\ln(1 - \mathrm{ST}_{it}) + \hat{\alpha}_i t)^2.$$

To ensure that the sell-through evolution is sufficiently well fitted through an exponential pattern, we calculated for each SKU the value $r_i^2 = 1 - \left[\sum_{t=1}^{W} (\ln(1 - \mathrm{ST}_{it}) + \alpha_i t)^2\right] / \left[\sum_{t=1}^{W} (\ln(1 - \mathrm{ST}_{it}) - \mu)^2\right]$, where μ is the average of $\ln(1 - \mathrm{ST}_{it})$ over the window. For a vast majority products, characterizing the sales diffusion as an exponential function of t closely fits the actual diffusion path: on average $r_i^2 = 0.93$.

Furthermore, it is worth highlighting that to make Eq. (4.6) consistent with the model of Sect. 4.3, it is necessary that α_i remains constant over the weeks in which the parameter is calculated. This is the case. Clearly, product characteristics (origin, family, color, cost, etc.) do not change. The only potential problem may come from changes in the pricing of the item. Fortunately, firm F has a policy of not running markdowns in season. To ensure that this policy was being followed in our data, we examined whether the same product had been sold at different prices in the same store (because different countries use different prices): We found that 67 % of store–SKU pairs had the same price recorded in the company's ticket database during the window over which we did the estimation; 89 % had a coefficient of variation of less than 1 %; and only 8 % of the pairs had a variation of more than 10 %. Even in those cases, there was no systematic markdown being applied: It seemed to be due to occasional discounting, e.g., through vouchers to the best customers or volume discounts, as prices seemed to vary erratically. This indicates that the price information includes factors unrelated to the scope of this paper that should not affect the computation of the speed of sales. As a result, when we include price effects later on, we only include the average sales price, instead of more granular price information.

Figure 4.2 shows the distribution of α_i, by region. For each region, we depict the empirical cumulative distribution function (c.d.f.) of speed of sales for all products being sourced for the same origin. We can already observe that regions with a shorter time-to-market have speed of sales distributions with higher averages and less dispersion. Of course, to perform a robust test of our hypothesis, we shall control for all the other factors that may influence α_i, which we do next.

4.4.5 Testing the Hypotheses

To validate our hypotheses, we test several models. We start with a one-factor model: Model I only includes the effect of design time but neglects time-to-market effects, and it can thus be used to evaluate the learning that takes place during the

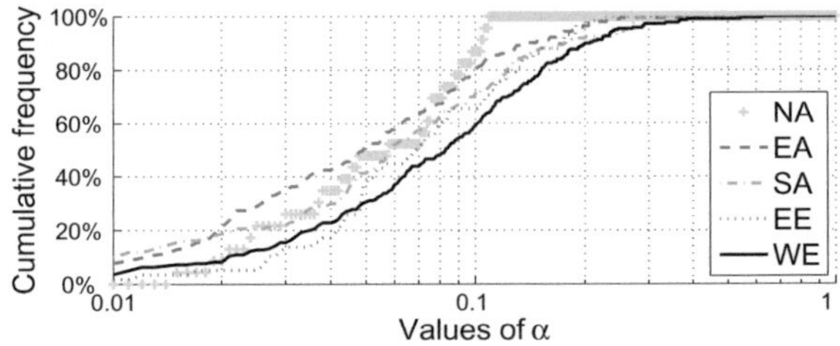

Fig. 4.2 C.d.f. of α_i for each origin

season, i.e., hypothesis H1. We then proceed to Model II, which considers design time and time-to-market variables. As pointed out in Sect. 4.4.3, these two variables exhibit strong collinearity, so the model also considers the interaction between design time and time-to-market. Furthermore, in all models, we control for all the observable product features, such as product family or color to take into account observable seasonal preferences. We also control for average sales price and cost, to take into account differences in product value. These two variables are somewhat correlated, and we observe that there is a relatively stable multiplier, i.e., the ratio between average price and cost, across all products. Finally, we also control by product purchase quantity to take into account possible effects of batch size on the speed of sales. Note nevertheless that our analytical model from Eq. (4.3) suggests that the sell-through is independent of the volume. Hence, although there is no conceptual support to consider that the speed of sales α_i may depend on the purchasing volume, we include it for completeness, and we find that it is not significant.

Our two models have a multiplicative form, i.e., we regress $\ln(\alpha_i)$ with the independent variables, rather than α_i. Indeed, from the tails of the distributions in Fig. 4.2, it seems more reasonable to use a multiplicative error model. In addition, we know that α_i must be positive, and a multiplicative form will avoid providing negative estimates. Finally, we should point out that additive models were also tested, but the fits are much better with multiplicative ones.

4.4.5.1 Model I

Model I is described by:

$$\begin{aligned}\ln(\alpha_i) = \beta_0 + \beta_D D_i + \beta_{\text{PRICE}} \cdot PRICE_i + \beta_{\text{COST}} \cdot COST_i \\ + \beta_Q \cdot Q_i + \sum_{c \in \text{Color}} \beta_c \cdot \chi_{c_i} + \sum_{c \in \text{Family}} \beta_f \cdot \chi_{f_i} + \epsilon_i\end{aligned} \tag{4.7}$$

D_i is the week in which the product was designed, as described in Sect. 4.3.2. Additionally, *Color* and *Family* are categorical variables for product color and

Table 4.2 OLS estimation for Model I (7)

	Estimate	Std. error	p value
Intercept	−3.58800	0.25410	2.0e−16***
D	0.02593	0.00323	5.3e−15***
Q	0.00002	0.00006	7.5e−01
PRICE	−0.03878	0.00782	9.2e−07***
COST	0.02884	0.01655	8.2e−02.
R^2	0.334		

Significance levels: '***': 0.001 and '.': 0.1

family, $PRICE_i$ is the average price recorded over all sales for the item in the store network, $COST_i$ is the product cost, and Q_i contains the product batch size.

The model parameters are estimated using OLS. For this model, there was no apparent relationship between D_i and the residual of the regression, so OLS was appropriate. In contrast, in later models, there is some heterogeneity across time-to-market and origins, so weighted least square (WLS) methods are used in those cases. Table 4.2 shows the results (color and family coefficients have been omitted).

The key insight from this model is that β_D is positive and significant, at the 0.001 level. This supports H1 and means that items designed later in the season will be more attractive and sell faster. The magnitude of this coefficient suggests that an item will sell 2.5 % faster than some other item designed 1 week before. Still, the low R^2 suggests that despite the significant average effect on sales, other product aspects may also be important to determine the speed of sales.

Moreover, one may wonder whether the learning takes place at the same pace during the season, e.g., whether most of the learning takes place in the beginning. This would suggest a nonlinear effect of design time on speed of sales. For this purpose, we tested a variation of this model with $\ln(D_i)$ and D_i^2, but the coefficients for these additional variables turned out to be non-significant.

Finally, it is interesting to report (from the complete model in the left) that items with lower price and higher cost sell faster, although the cost relationship is significant at the 10 % level only. On the one hand, this reflects the price sensitivity of customers. On the other, it suggests that customers also react to product cost: In other words, products that cost more, i.e., require more expensive fabrics or more labor, offer higher value to consumers.

4.4.5.2 Model II

Model I only included the effect of design time on sales. This implies that there is no difference in treatment between products with different time-to-market. For instance, a product designed in week 25 and made in West Europe is introduced into the stores on week 27, while another made in South Asia arrives in week 37. We thus include in Model II.a both variables D_i and T_i. In addition, the interaction term $D_i \times T_i$ is also incorporated, to identify whether the learning effect during the

Table 4.3 WLS estimation for Model II.a (8) on the left and Model II.b (9) on the right

	Estimate	Std. error	p value	Estimate	Std. error	p value
Intercept	−3.21100	0.35830	2.0e−16***	−2.82300	0.33190	2.0e−16***
D	0.01552	0.00729	3.4e−02*	0.00626	0.00622	3.2e−01
T	−0.03451	0.01988	8.3e−02.			
D × T	0.00104	0.00074	1.6e−01			
OFFSHORE				−0.94390	0.25600	2.5e−04***
D × OFFSHORE				0.02771	0.00937	3.2e−03**
Q	0.00002	0.00006	7.1e−01	−0.00001	0.00006	8.4e−01
PRICE	−0.03875	0.00796	1.4e−06***	−0.03832	0.00776	1.0e−06***
COST	0.02991	0.01671	7.4e−02.	0.02829	0.01628	8.3e−02.
R^2	0.344			0.354		

Significance levels: '***': 0.001; '**': 0.01; '*': 0.05; and '.': 0.1

season (seen in Model I) is stronger or weaker depending on the speed of execution. It is worth pointing out that there was some heteroskedasticity across regions, so we used WLS to estimate the parameters, with the weights being equals to the variance of the residuals obtained from OLS estimation. Interestingly, the variance of the residuals is higher for higher time-to-market, i.e., there is more uncertainty in the success rate of products made off shore.

$$\ln(\alpha_i) = \beta_0 + \beta_D D_i + \beta_T T_i + \beta_{D\times T} D_i T_i + \beta_{\text{PRICE}} \cdot \text{PRICE}_i + \beta_{\text{COST}} \cdot \text{COST}_i \\ + \beta_Q \cdot Q_i + \sum_{c\in\text{Color}} \beta_c \cdot \chi_{c_i} + \sum_{c\in\text{Family}} \beta_f \cdot \chi_{f_i} + \epsilon_i \tag{4.8}$$

The results are shown in the left of Table 4.3. Since D_i, T_i and $D_i \times T_i$ are correlated, we see that the significance of all these variables is reduced compared to Model I. In particular, part of the learning effect identified in Model I is taken now by variable T_i (this effect is negative because items made in regions with short time-to-market are designed later). In addition, the goodness of fit is not improved compared to Model I, as R^2 is only marginally higher. As a result, the model in (4.8) does not seem to capture correctly the joint effects of D_i and T_i. Hence, we develop Model II.b, a simpler alternative to (4.8) in (4.9), by including the variable $OFFSHORE_i$ instead of T_i (which was numerical but limited to values of 2, 4, 12, or 16 weeks). Recall that $OFFSHORE_i$ is equal to one when the item is sourced in Asia (East and South) and zero otherwise:

$$\ln(\alpha_i) = \beta_0 + \beta_D D_i + \beta_{\text{OFFSHORE}} \text{OFFSHORE}_i + \beta_{D\times\text{OFFSHORE}} D_i \text{OFFSHORE}_i \\ + \beta_{\text{PRICE}} \cdot \text{PRICE}_i + \beta_{\text{COST}} \cdot \text{COST}_i + \beta_Q \cdot Q_i \\ + \sum_{c\in\text{Color}} \beta_c \cdot \chi_{c_i} + \sum_{c\in\text{Family}} \beta_f \cdot \chi_{f_i} + \epsilon_i \tag{4.9}$$

The right of Table 4.3 contains the results of this alternative Model II.b. First, we see that the model offers a slightly higher goodness of fit, because T_i and D_i are correlated so the informativeness of T_i beyond D_i is not that high. To obtain a higher R^2, one could consider the categorical variables $ORIGIN_i$ instead of T_i, but then there would be few observations in East Europe and North Africa, so we focused on this more robust specification.

The learning effect during the season is now weakened, and most of this effect is taken by $OFFSHORE_i$: Items made in Asia have lower speed of sales. But this does not mean that there is no learning. Indeed, the positive interaction term implies that the learning effect is different depending on the origin. It is significantly larger for offshore origins compared to nearshore ones. This indicates that as season progresses, the success of products from offshore countries improves faster than those from nearshore ones. In other words, offshore products benefit a lot by delaying the design time, which suggests that 'blind' designs made very early carry high design risk and hence will on average be less successful, while 'informed' designs made after the beginning of the season perform much better. In contrast, the learning for nearshore products is small. We interpret that this is due to having most of these products designed in season, so that most of the learning from early sales is already present. This suggests that learning is much higher in the first half of the season (weeks 20–35), while it is limited in the second half (weeks 35–50).

To evaluate the absolute effect of short time-to-market on speed of sales, consider for instance a product designed in week 20: The difference between the 'predicted' average log speed of sales for offshore products ($-2.79 - 0.94 + 0.027 \times 20$ weeks) and nearshore ones (-2.79) is quite important, about 49 % higher for nearshore items. This difference is reduced as design time increases: By week 25, it is 30 % higher; by week 30, only 14 %. For a product designed in week 35, the difference disappears, but this is already too late in the season to order items from Asia. As a result, although short time-to-market remains valuable, within the range of the data (T_i being smaller than 35), H2 is not supported because by the middle of the season the difference between offshore and nearshore becomes insignificant.

In contrast, we find support for H1 and H3. As the season progresses, the average success rates increase, so postponing design is valuable; and the difference of success rates across offshore and nearshore origins tends to vanish gradually as design time increases. In other words, the learning is stronger for products made in Asia, which are designed very early. Indeed, products from faraway origins are mostly 'wild guesses' of season trends prior to the season, and one would expect much lower average success rates, as we observe. As the season advances and customer preferences start to be observed, designs from offshore origins are much closer to what the market is expecting but, still, long time-to-market makes these products less successful compared to those sourced near shore. Eventually, designs made well into the season are able to have designs closer to what the market expects and thus have success rates comparable to nearshored items. As discussed in Sect. 4.3.2, this seems to indicate that trends can fashion trends are persistent, as opposed to short-lived. They are revealed by early sales, and by week 35, there is no significant advantage of a short

Table 4.4 WLS estimation to explain the sell-through after 4 weeks (ST_4, top) and 8 weeks (ST_8, bottom)

	Estimate	Std. error	*p* value	Estimate	Std. error	*p* value
$\ln(ST_4)$						
Intercept	−2.55500	0.48320	1.7e−07***	−2.18600	0.43710	7.5e–07***
D	0.02773	0.01089	1.1e−02*	0.02044	0.00905	2.4e–02*
T	−0.14910	0.02849	2.3e−07***			
D × T	0.00625	0.00104	3.6e−09***			
OFFSHORE				−2.66500	0.35070	1.1e–13***
D × OFFSHORE				0.10050	0.01269	1.2e–14***
Q	−0.00002	0.00007	7.4e−01	−0.00008	0.00008	2.9e–01
PRICE	−0.03871	0.01083	3.8e−04***	−0.03209	0.01007	1.5e–03**
COST	0.04648	0.02225	3.7e−02*	0.03333	0.02084	1.1e−01
R^2	0.443			0.480		
$\ln(ST_8)$						
Intercept	−1.48100	0.45670	1.3e−03**	−1.03200	0.42100	1.5e–02*
D	0.02539	0.01169	3.0e−02*	0.01639	0.01018	1.1e−01
T	−0.10010	0.02650	1.8e−04***			
D × T	0.00409	0.00099	3.9e−05***			
OFFSHORE				−1.92200	0.34350	3.5e–08***
D × OFFSHORE				0.07059	0.01256	3.0e–08***
Q	−0.00002	0.00006	7.9e−01	−0.00003	0.00006	5.9e–01
Price	−0.03962	0.00904	1.4e−05***	−0.03995	0.00869	5.3e–06***
Cost	0.04483	0.01825	1.4e−02*	0.04385	0.01760	1.3e−02*
R^2	0.468			0.496		

Significance levels: '***': 0.001; '**': 0.01; '*': 0.05; and '.': 0.1

time-to-market. Hence, the value of QR resides in being able to defer the design time so as to take advantage of trend information.

4.4.6 Alternative Measures of Success

So far we have focused our attention on speed of sales. To ensure the robustness of our conclusions, we have tested our explanatory models with other measures of product success. Namely, we have regressed the traditional metric of sell-through after a fixed number of weeks, with all the variables present in our study. Table 4.4 contains the results of Models II.a and II.b applied to the logarithm of sell-through rates after 4 and 8 weeks. The conclusions remain unchanged: being able to design later and having a short time-to-market both increase the success rate, and the learning effect is more pronounced for offshore sources.

4.4.7 Sourcing Choices

Based on the process that F uses to allocate products to origins and suppliers, we have assumed throughout our analysis that the sourcing choice is exogenous. One potential concern is that such choice may be endogenous which would clearly impact our results. In fact, endogeneity would actually strengthen our conclusions, although the actual estimates would be different.

Specifically, one could imagine that the firm chooses where to produce a product based on its prospects. Unfortunately, we have no way of identifying such behavior with our data, because product forecasts or other qualitative assessments are not available. But if product-source allocation was indeed endogenous, Fisher (1997) describes how to so: Products with higher a priori uncertainty should be sent nearshore. That is, the company may source riskier products from nearshore origins and more basic risk-free products from offshore origins, and this should endogenously affect speed of sales. As a result, one should expect a lower or more uncertain success rate of products with short time-to-market. We find the opposite, as discussed before Eq. (4.8). Thus, our results are robust in sign and significance to this potential bias, although the absolute value of the estimator may be biased.

4.5 Discussion

Our paper presents an empirical study of the value of quick response. We first build an analytical model where sales are related to inventory availability. We use this model to explain the dynamics of sales in our data and to construct the success indicator of a product: We define speed of sales as the share of inventory that is sold per time unit. Thus, more successful products sell faster, generally more but not necessarily (because they may stay in the store for a shorter period). We then compute the speed of sales of all the products in the sample and study how it depends on production choices, specifically the design time and the time-to-market (long if made off shore, short near shore).

Several conclusions can be extracted, which are new to the literature. First, our results confirm that QR practices can help a fashion company to increase the speed of sales. In our case, the ability of postponing design provides a significantly higher speed of sales (H1). Second, we observe that shorter time-to-market in itself is also valuable, although only significant early in the season (H2). Third, the improvement obtained by postponing design is more pronounced when time-to-market is longer, hence for offshore origins compared to nearshore ones (H3). This suggests that nearshore production is generally able to produce faster-selling products, but offshore production after early seasonal trends are observed can perform well too. In terms of magnitude, from our Model II, we can state that products made in West Europe, East Europe, and North Africa in the middle of the season (week 25) have a speed of sales that is about 30 % higher ($e^{0.94-0.027\times 25}$) than those made in East Asia and South Asia.

Our work has important implications. Namely, our results suggest that choosing a supplier in a particular region does have an impact on the speed at which sales are realized. This can be used for deciding where to produce each of the items to be introduced in the store in a given time. For example, consider that a new item (e.g., within the T-shirt category) needs to be introduced in week 30 of a given year. The item can be designed and procured from East Asia in week 14 of the year (4 months before the introduction date), or in South Asia in week 18 (3 months before), or nearshore in week 26 (1 month before). Each choice determines a different level of speed of sales and uncertainty. It will also impact costs, which may be higher or lower depending on the opportunity cost of shelf space (which is occupied for a longer time with offshore production) and the procurement cost associated with the suppliers. This is a critical trade-off for the industry. As stated in H&M (2007), different items require different lead times: 'The time from an order being placed until the items are in the store may be anything from a few weeks up to six months. The best lead time will vary for a particular situation. For high-volume fashion basics and children's wear it is advantageous to place orders further in advance. In contrast, trendier garments in smaller volumes have to be in the stores much quicker.' Our study provides estimates for the drivers of these decisions.

Our study has several limitations that should be explored in further research. First, we work with a sample of one season in one company. The company is a European retailer that is representative of the best practices of fast fashion. It would be interesting to conduct similar studies in other fast fashion firms and a longer time span. Ideally, it could combine with controlled field experiments, as in Caro and Gallien (2010). Second, it would be valuable to validate and compare the results with other similar data sets, for example, for high-end fashion products, where brands try to create trends, rather than following them; or in other markets such as Asia or the USA. One should nevertheless note that this type of data is quite sensitive and firms tend to be reluctant to share it for research purposes. Third, more analytical work that explicitly model demand dynamics related to inventory availability seems necessary, as it seems to fit the data much better than models with exogenous stochastic demand.

Acknowledgements V. Martínez-de-Albéniz's research was supported in part by the European Research Council—ref. ERC-2011-StG 283300-REACTOPS and by the Spanish Ministry of Economics and Competitiveness (Ministerio de Economa y Competitividad)—refs. ECO2011-29536 and ECO2014-59998-P.

References

Amihud Y, Medenelson H (1989) Inventory behaviour and market power: an empirical investigation. Int J Ind Organ 7(2):269–280

Balakrishnan A, Pangburn M, Stavrulaki E (2004) "Stack them high, let'em fly": lot-sizing policies when inventories stimulate demand. Manage Sci 50(5):630–644
Baron O, Berman O, Perry D (2011) Shelf space management when demand depends on the inventory level. Prod Oper Manage 20(5):714–726
Berfield S, Baigorri M (2013) Zara's fast-fashion edge. Business week, Nov 14 (online)
Bernstein F, Martínez-de-Albéniz V (2014) Dynamic product rotation in the presence of strategic customers. Manage Sci Forthcoming
Butler S (2013) Inditex: Spain's fashion powerhouse you've probably never heard of. The Guardian, Dec 15 (online)
Cachon G, Swinney R (2011) The value of fast fashion: quick response, enhanced design, and strategic consumer behavior. Manage Sci 57(4):778–795
Cachon GP, Terwiesch C (2009) Matching supply with demand: an introduction to operations management. McGraw-Hill/Irwin, New York, USA
Caldentey C, Caro F (2010) Dynamic assortment planning. Working paper, UCLA
Caro F (2011) ZARA: staying fast and fresh. Technical report, UCLA Anderson School of Management case 612-006-1
Caro F, Gallien J (2007) Dynamic assortment with demand learning for seasonal consumer goods. Manage Sci 53(2):276–292
Caro F, Gallien J (2010) Inventory management of a fast-fashion retail network. Oper Res 58 (2):257–273
Caro F, Gallien J, Daz M, Garca J, Corredoira J, Montes M, Ramos J, Correa J (2010) Zara uses operations research to reengineer its global distribution process. Interfaces 40(1):71–84
Caro F, Martínez-de-Albéniz V (2013) Operations management in apparel retailing: processes, frameworks and optimization. BEIO, Boletn de Estadstica e Investigación Operativa 29 (2):103–116
Caro F, Martínez-de-Albéniz V (2015) Fast fashion: business model overview and research opportunities. In: Agrawal N, Smith SA (eds) Retail supply chain management: quantitative models and empirical studies, 2nd edn, ed. Springer, New York
Caro F, Martínez-de Albéniz V, Rusmevichientong P (2014) The assortment packing problem: multiperiod assortment planning for short-lived products. Manage Sci 60(11):2701–2721
Chen H, Frank M, Wu O (2005) What actually happened to the inventories of American companies between 1981 and 2000? Manage Sci 51(7):1015–1031
Christopher M, Lowson R, Peck H (2004) Creating agile supply chains in the fashion industry. Int J Retail Distrib Manage 32(8):367–376
Çnar E, Martínez-de-Albéniz V (2013) A closed-loop approach to dynamic assortment planning. Working paper, IESE Business School
Corstjens M, Doyle P (1981) A model for optimizing retail space allocations. Manage Sci 27 (7):822–833
Curhan RC (1973) Shelf space allocation and profit maximization in mass retailing. J Mark 37 (3):54–60
El País (2008) Estrenar ropa dura segundos. El País, Jan 2 (online)
El País (2011) Inditex escapa a la crisis. El País, Jan 16 (online)
Eppen G, Iyer A (1997) Backup agreements in fashion buying—the value of upstream flexibility. Manage Sci 43(11):1469–1484
Feitzinger E, Lee H (1997) Mass customization at Hewlett-Packard: the power of postponement. Harvard Bus Rev 117(1):116–121
Ferdows K, Machuca JAD, Lewis M (2002) Zara. Technical report, ECCH case 603-002-1
Fisher M (2009) OR FORUM–rocket science retailing: the 2006 Philip McCord Morse Lecture. Oper Res 57(3):527–540
Fisher ML (1997) What is the right supply chain for your product? Harvard Bus Rev 75:105–117
Fisher ML, Rajaram K, Raman A (2001) Optimizing inventory replenishment of retail fashion products. Manuf Serv Oper Manage 3(3):230–241
Fisher ML, Raman A (1996) Reducing the cost of demand uncertainty through accurate response to early sales. Oper Res 44(1):87–99

Foroohar R (2006) A new fashion frontier. Newsweek International, Mar 19 (online)
Gaur V, Fisher M, Raman A (2005) An econometric analysis of inventory turnover performance in retail services. Manage Sci 51(2):181–194
Gaur V, Honhon D (2006) Assortment planning and inventory decisions under a locational choice model. Manage Sci 52(10):1528–1543
Ghemawat P, Nueno JL (2003) ZARA: fast fashion. Technical report, Harvard Business School Multimedia Case 9-703-416
Hammond JH, Kelly MG (1990) Quick response in the apparel industry. Technical report, Harvard Business School Note 9-690-038
Heath DC, Jackson PL (1994) Modeling the evolution of demand forecasts with application to safety stock analysis in production/distribution systems. IIE Trans 26(3):17–30
H&M (2007) Annual report
Iyer AV, Bergen ME (1997) Quick response in manufacturer-retailer channels. Manage Sci 43 (4):559–570
Keeley G, Clark A (2008) Zara overtakes gap to become world's largest clothing retailer. The Guardian, Aug 11 (online)
Kesavan S, Gaur V, Raman A (2010) Do inventory and gross margin data improve sales forecasts for U.S. public retailers? Manage Sci 56(9):1519–1533
Lee H, Tang C (1997) Modelling the costs and benefits of delayed product differentiation. Manage Sci 43(1):40–53
Martínez-de-Albéniz V, Roels G (2011) Competing for shelf space. Prod Oper Manage 20(1):32–46
McAfee A, Dessain V, Sjöman A (2004) ZARA: IT for fast fashion. Technical report, Harvard Business School case 9-604-081
Musalem A, Olivares M, Bradlow E, Terwiesch C, Corsten D (2010) Structural estimation of the effect of out-of-stocks. Manage Sci 56(7):1180–1197
Olivares M, Cachon G (2009) Competing Retailers and Inventory: an empirical investigation of general motors' dealerships in isolated U.S. markets. Manage Sci 55(9):1586–1604
Rumyantsev S, Netessine S (2007) What can be learned from classical inventory models? A cross-industry exploratory investigation. Manuf Serv Oper Manage 9(4):409–429
Smith S, Achabal D (1998) Clearance pricing and inventory policies for retail chains. Manage Sci 44(3):285–300
Song J-S, Zipkin PH (2012) Newsvendor problems with sequentially revealed demand information. Naval Res Logistics 59(8):601–612
Ton Z, Raman A (2010) The effect of product variety and inventory levels on retail store sales: a longitudinal study. Prod Oper Manage 19(5):546–560
Wang Y, Gerchak Y (2001) Supply chain coordination when demand is shelf-space dependent. Manuf Serv Oper Manage 3(1):82–87

Chapter 5
Mixed Channels for Apparel Sales

Xiaowei Zhu, Samar K. Mukhopadhyay and Xiaohang Yue

Abstract We study a clothing company which sells clothes through a traditional retail store and also sells clothes to the end customers through a direct online channel. The retail store offers some value-added services to the customers and is also given full authority to make pricing decision. Under this scenario, we present a mixed channel model to obtain optimal pricing decisions by both parties. We explore the amount of value adding by the retailer and determine the manufacturer's optimal wholesale price to the retailer. Our model considers information asymmetry where the manufacturer has incomplete information about the retailer's cost of adding value. We obtain closed-form contracts with incomplete information and compare them with those when complete channel coordination is present. We develop a number of managerial guidelines for clothing companies and identify future research topics.

Keywords Mixed channel · Multi-channel retailing · Apparel retailing · Information asymmetry · Channel coordination

---

X. Zhu (✉)
College of Business and Public Affairs, West Chester University of Pennsylvania,
West Chester, PA 19383, USA
e-mail: xzhu@wcupa.edu

S.K. Mukhopadhyay
Graduate School of Business, Sungkyunkwan University, Jongno-Gu,
Seoul 110-745, South Korea
e-mail: samar@skku.edu

X. Yue
Sheldon B. Lubar School of Business, University of Wisconsin-Milwaukee,
P.O. Box 742, Milwaukee, WI 53201, USA
e-mail: xyue@uwm.edu

T.-M. Choi (ed.), *Analytical Modeling Research in Fashion Business*,
Springer Series in Fashion Business, DOI 10.1007/978-981-10-1014-9_5

5.1 Introduction and Literature Survey

More and more companies are opening direct channels for selling fashion clothes to customers, which are in addition to their brick-and-mortar retailer channel. The sales associates at retail store engage in activities to build brand awareness, gather market information, and provide customer support including providing the opportunity of returns. More importantly, sales associates can help each individual customer with their unique buying behaviors and preferences. Through a face-to-face communication, the experienced sales associates can better understand their customers' need and lifestyle and then suggest the most suitable style, color, and size to the customer. Online customers may buy the wrong style, color, or size without being aware that different style, color, or the size up or down may be more attractive. The Internet revolution and emerging technology play a pivotal role in the development of this new business model. From the online channel, the firm can directly capture customers' orders, inform customers of the product variety, have control over the goods distribution and pricing, and, being closer to customers, can have a greater understanding of the customers' preferences. Many companies use the multi-channel or omni-channel, which is the use of a variety of channels in a customer's shopping experience including research before a purchase. The channels might include retail stores, online stores, mobile stores, telephone sales, TV sales, and others.

Many existing literature, such as Balasubramanian (1998) and Levary and Mathieu (2000), suggest that a mixed channel structure could work well. Nike and Victoria's Secret (Collinger 1998), Old Navy, Gap, Banana Republic, Forever 21, Burberry, Louis Vuitton, and others pursue a mixed channel strategy. One of the common questions for the manufacturer is how to balance two channels and reduce the channel competition. Channel conflict (Tsay and Agrawal 2004) is the biggest deterrent for the manufacturer to go ahead with the mixed channel business model. As Frazier (1999) showed, mixed channel would increase revenue, but would lead to decreased support from the channel partners. In fact, this led to some retailers actually taking action against the manufacturers who opened a direct channel in competition with them. Tsay et al. (1999) and Frazier (1999) survey channel structure and incentive design for performance enhancement, but not channel conflict. Cohen et al. (1995) study an intermediary who perform specific value-added functions and get compensated for this service by the manufacturer or distributors by a side payment. A mixed channel strategy in products that do not provide large value is studied by Chiang et al. (2003) who show that adding a direct channel can mitigate the profit loss. Yao and Liu (2003) study diffusion of customer between two channels and find that, under certain conditions, both channels would enjoy stable demand. Beck and Rygl (2015) discuss the development of multi-channel conceptually and subsequently discuss existing research in this multi-channel retailing. Verhoef et al. (2015) propose a categorization of multi-, cross-, and omni-Channel retailing for retailers and retailing by means of a literature review, a taxonomy of multiple channel retailing, a literature classification table, and by way of illustration, a mobile Click and Collect shop. Kozlenkova et al. (2015) review the marketing and SCM literatures for the channel structure. Hübner et al. (2015) have questioned 43 executives from 33 different European-based leading companies in multi-channel retailing to study the operations structures of multiple channels, including the network design, inventory management, warehouse operations, and capacity management.

There are also some brands do not sell online or mainly depend on one channel due to the brand strategy. Aside from cosmetics and fragrances, labels such as Chanel, Céline, Hermes, and Dior require the customers physically go to a store to purchase most if not all of their clothes and handbags (Mau 2014). Mau (2014) reports that Bruno Pavlovsky, Chanel's president of global fashion, explained Chanel's sale strategy "Fashion is about clothing, and clothing you need to see, to feel, to understand" and also addressed the company's digital initiatives were designed "more to bring the customers to the boutique." From the above discussion, we see that the brand strategy is one of the important factors when designing the sale channels. In this research, we will discuss a mixed channel structure, an online channel, and retailer channel, based on different brand strategies, and make recommendations.

In this paper, we address the mixed channel strategy where the manufacturer can sell product directly online, and let the retailer offer some value-added services to the product before selling to the final customer (Fay 1999). Though channel structures have been extensively researched in the literature, relatively few have studied mixed channel with value-added retailer. We study three business models where the retailer–manufacturer conflict is alleviated by a contract, including (5.1) a base case, for benchmarking purpose, where the channel is integrated and a joint profit function is maximized; (5.2) a case where the channel partners are separate, but they share full information with each other; and (5.3) a more general case where there is information asymmetry in the channel. Under the information asymmetry, one partner offers a lumpsum side payment to the other to alleviate channel conflict. Here, we briefly review the business contract design under asymmetric information and/or information sharing. Desiraju and Moorthy (1997) study the case of information asymmetry about a price and service-sensitive demand curve. They show that coordination can be achieved by the requirement of service performance. Gan et al. (2003) find that information asymmetry about manufacturer's production cost does not necessarily cause inefficiency in the supply chain. Value of information in a capacitated supply chain is derived by Gavirneni et al. (1999). Lee et al. (2000) show that, with a demand process correlated over time, it could be worthwhile to share information about the demand. Corbett and de Groot (2000) derive optimal discount policy for both full and incomplete information cases. Corbett et al. (2004) study different types of contracts to coordinate the supply chain for both complete information and asymmetric information asymmetry. Aviv (2002) study the value of information sharing in the collaborative forecasting process. Gavirneni (2002) showed that the total supply chain cost can be lowered by information sharing. Ha (2001) finds that in case of private information, optimal order quantity is smaller and optimal selling price is higher than for the case with complete information.

Viswanathan (2000) studies the mixed channel issue from the product differentiation point of view and concludes that the more different the product is in the two channels, the more the benefit the channels have.

We will introduce our business scenario and the mixed channel model in Sect. 5.2. In Sect. 5.3, we derive the optimal contracts for the case of complete information and asymmetric information in addition to a base case. In Sect. 5.4, we compare three business cases through mathematic analysis and numerical experimentation to get several managerial insights. Section 5.5 will conclude the paper with some further research ideas.

5.2 The Model Scenario

We consider a mixed channel supply chain where a manufacturer sells clothes to the end customers through a retailer and this manufacturer also sells clothes directly to the same customer pool (see Fig. 5.1). At the retail store, the sales associates can offer some value-added services to the end customers. We list the symbols in Table 5.1.

The decision variables in our model are p_1, w, and L for the manufacturer and p_2 and v for the retailer, each maximizing their own profit functions.

5.2.1 Demand Functions

To maintain analytical tractability, we use the following demand functions. For detailed discussion of these demand functions, please refer to Mukhopadhyay et al. (2008).

$$\text{Direct channel: } d_1 = a - p_1 + r(p_2 - v) \tag{5.1}$$

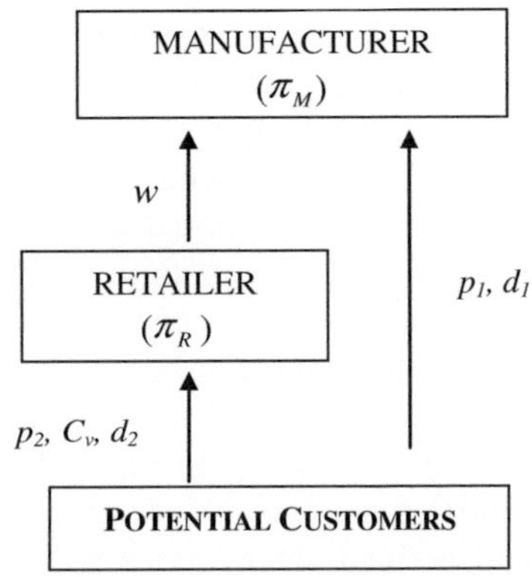

Fig. 5.1 Mixed channel for fashion clothes sale

Table 5.1 Symbols used in the article

Symbol	Stand(s) for
a	Base demand. Common knowledge
d_1 and d_2	Demand function for online and retail channel, respectively
p_1	Retail price charged at online channel. Decision variable for manufacturer
p_2	Retail price charged at retail channel. Decision variable for retailer
w	Wholesale price charged by the manufacturer to the retailer. Decision variable for manufacturer
r	Migration effectiveness between channels. Common knowledge. $0 < r < 1$
v	Value added to the basic product by the sales associate at the retail store. Decision variable for retailer
η	Efficiency parameter for the retailer's value-added cost. Retailer's private information in asymmetric information case
c_v	Cost function for value-added service
L	Side payment, from manufacturer to the retailer. Decision variable for manufacturer under asymmetric information case
π_R and π_M	Retailer's and manufacturer's profit, respectively
$\pi_{\overline{R}}$ and $\pi_{\overline{M}}$	Retailer's and manufacturer's reservation profit, respectively
(*F*), (*A*), (*I*)	Full information, integrated case, and asymmetric information, respectively

$$\text{The retailer channel:}\, d_2 = a - (p_2 - v) + r p_1 \tag{5.2}$$

Let the equilibrium demands be d_1 for the direct online channel and d_2 for the retailer channel. Let p_1 be the price of the basic product charged by the manufacturer in the direct online channel. Let v be the value added to the basic product by the sales associate at the retail store where the sale prices is p_2. When both the channels are operating at the same time, customers would make a purchase decision by considering the two prices p_1 and p_2 and also the value added v by the sales associate at retail store. d_1 and d_2, therefore, would be functions of p_1, p_2, and v. We assume $p_2 > p_1$. There will be migration of customers from one channel to another. r is the migration effectiveness. We assume that $0 < r < 1$, so that own channel effects are greater than cross-channel effects. a and r are assumed to be common knowledge.

5.2.2 *Cost Function for Value Adding*

The cost for the sales associate to offer a value-added service is c_v per unit. We use a quadratic cost function for the sales associate to offer a value-added service. Specifically, we use the functional form:

$$c_v = \eta \frac{v^2}{2} \tag{5.3}$$

The quadratic cost function can capture the phenomenon that adding a large quantum of value is proportionately more costly than adding minimal amount of value. η is an efficiency parameter for the retailer to offer value-added services. Small value of η means high efficiency in offering value-added services. From the Eq. (5.3), we can see that the more efficient (or smaller η) is the retail store, the less costs (or smaller c_v) the retail store to offer the value-added service. η is retailer's private information under asymmetric information case. In other words, the manufacturer does not know how efficient the retail store can offer the value-added service. In this paper, we consider contracts under two information structures. In full information scenario, the retailer shares her private information η to the manufacturer. In asymmetric information scenario, the retailer will not share her private information η to the manufacturer and so manufacturer does not know η. The manufacturer holds a prior cumulative distribution $F(\eta)$ with density function $f(\eta)$ about η.

5.2.3 *Profit Functions*

We use the following profit functions for the retailer and manufacturer:

$$\text{The retailer's profit function: } \pi_R = (p_2 - w - c_v)d_2 + L \tag{5.4}$$

$$\text{And the manufacturer's profit function: } \pi_M = p_1 d_1 + w d_2 - L \tag{5.5}$$

where d_1 and d_2 are given by Eqs. (5.1) and (5.2). w is the wholesale price charged by the manufacturer to the retailer. Each of the retailer and the manufacturer has a reservation profit level $\pi_{\overline{R}}$ and $\pi_{\overline{M}}$, respectively. If one side or both sides' profits drop below the reservation profit(s), there will be no contract or no trade between the retailer and manufacturer.

5.3 The Contracts

We will study three kinds of contract: contract under Channel integration (*I*), under full information (*F*), and under asymmetric information (*A*). Under (*I*), the contract is designed by maximizing the total profit of manufacturer and retailer and taking η as common knowledge. Under (*F*), the manufacturer knows the retailer's η and designs the contract, taking η as common knowledge. Each of the manufacture and retailer maximizes his or her own profit, respectively. Under (*A*), each of the manufacture and retailer also maximizes his or her own profit, respectively. But the manufacturer does not know the retailer's η. The manufacturer needs to design the contract using the prior density function $f(\eta)$ and cumulative distribution $F(\eta)$ of η.

5.3.1 *Contract Under Channel Integration* (*I*)

In this case, the two channels are vertically integrated where the manufacturer and retailer behave like a single firm, optimizing the total jointed channel profit. It is expected that the channel profit would be highest under this scenario. This case is the first best case and will be used as a base case for comparison.

$$\pi^I = p_1 d_1 + [p_2 - c_v] d_2 \tag{5.6}$$

The optimal online prices p_1, retail store price p_2, and value-added level v can be obtained by taking the first-order condition of the total profit with respect to p_1, p_2, and v and then solve them simultaneously. The following Proposition 1 gives the optimal contract values for the integration case (I).

Proposition 1

(a) *The optimal contract under channel integration is given by:*

$$p_1^I = \frac{a}{2(1-r)} \quad p_2^I = \frac{a}{2(1-r)} + \frac{3}{4\eta} \quad v^I = \frac{1}{\eta}$$

$$N^I = \frac{1}{-2a + 2\sqrt{4\pi_{\overline{M}} + 4\pi_{\overline{R}} - a^2\frac{1+r}{1-r}}} \quad \text{where } \pi_{\overline{M}} + \pi_{\overline{R}} \geq \frac{a^2(1+r)}{4(1-r)}$$

(b) *The optimal profits for the whole channel, the retailer, and the manufacturer under channel integration are given by:*

$$\text{Total profit: } \pi^I = \frac{a}{4\eta} + \frac{a^2}{2(1-r)} + \frac{1}{16\eta^2}$$

If the retailer earns her reservation profit: $\pi_R^I = \pi_{\overline{R}}$, the manufacturer will get

$$\pi_M^I = \begin{Bmatrix} \frac{a}{4\eta} + \frac{1}{16\eta^2} + \frac{a^2}{2(1-r)} - \pi_{\overline{R}} & \underline{\eta} \leq \eta \leq N \\ \pi_{\overline{M}} & N \leq \eta \leq \overline{\eta} \end{Bmatrix}$$

From the Proposition 1(b), we get the total profit for the manufacturer and retailer. The manufacturer and retailer might need to decide how this total profit be divided between the two. The factors that affect the profit divisions include the potential customers' shopping preference (online or at the store), brand strategy, and bargaining power between the manufacturer and retailer. In Proposition 1(b), we give a possible way of dividing the total profit where the retailer is given $\pi_{\overline{R}}$ to ensure her participation and the manufacturer, therefore, receives the remaining profit $\pi^I - \pi_{\overline{R}}$. A contract like this is proposed by Corbett et al. (2004).

5.3.2 *Contracts Under Full Information* (F)

Under full information (*F*), the retailer shares her private information η with the manufacturer. The moves of manufacturer and retailer follow a Stackelberg-type game: The manufacturer acts as the leader, announcing the p_1 and w first; the retailer acts as the follower, announcing the p_2 and v after that.

This Stackelberg game can be solved backward. Before the first stage of the game where the manufacturer announces his decision variables, the retailer's best response function can be calculated by maximizing her profit π_R with respect to her decision variables, namely p_2 and v. Equation (5.7) gives the retailer's best response function, as functions of p_1 and w.

$$p_2^r = \frac{3}{4\eta} + \frac{w}{2} + \frac{a}{2} + \frac{p_1 r}{2} \quad v^r = \frac{1}{\eta} \tag{5.7}$$

In Stage 1 of the game, the manufacturer derives his optimal p_1 and w by maximizing his profit function π_M, given in Eq. (5.5). The manufacturer also needs to substitute p_2 and v with the retailer's best response functions p_2^r and v^r in Eq. (5.7). Then, the manufacturer's profit function only includes his decision variables p_1, w, and other parameters (common knowledge). Using the first-order conditions, we obtain the manufacturer's optimal policies as follows:

$$p_1^F = \frac{a}{2(1-r)} \quad w^F = \frac{a}{2(1-r)} + \frac{1}{4\eta} \tag{5.8}$$

In Stage 2 of the game, the retailer uses the manufacturer's policy announcement given in Eq. (5.8) and maximizes her profit function, given in Eq. (5.5), to obtain her optimal policies as follows:

$$p_2^F = \frac{(3-r)a}{4(1-r)} + \frac{7}{8\eta} \quad v^F = \frac{1}{\eta} \tag{5.9}$$

The following Proposition 2 gives the optimal contract values for the full information case (*F*).

Proposition 2

(a) *The optimal contract under full information is given by*:

$$p_1^F = \frac{a}{2(1-r)} \quad w^F = \frac{a}{2(1-r)} + \frac{1}{4\eta}$$
$$p_2^F = \frac{(3-r)a}{4(1-r)} + \frac{7}{8\eta} \quad v^F = \frac{1}{\eta}$$

(b) *The optimal profits for the whole channel, retailer, and* manufacturer *under full information are given by:*

$$\text{Total profit: } \pi^F = \frac{3a}{16\eta} + \frac{(7+r)a^2}{16(1-r)} + \frac{3}{64\eta^2},$$

$$\pi_M^F = \frac{a}{8\eta} + \frac{(3+r)a^2}{8(1-r)} + \frac{1}{32\eta^2}, \quad \pi_R^F = \frac{a}{16\eta} + \frac{a^2}{16} + \frac{1}{64\eta^2}$$

5.3.3 *Contracts Under Asymmetric Information* (A)

Under asymmetric information (*A*) case, the retailer will not share her private information η with the manufacturer. The retailer keeps her cost efficiency parameter η as a private information. The moves of manufacturer and retailer also follow a Stackelberg-type game: The manufacturer acts as the leader, announcing the p_1 and w first; the retailer acts as the follower, announcing the p_2 and v after that. In this Stackelberg game, when the manufacturer needs η to decide his optimal decision variables, he will use the prior density function $f(\eta)$ and cumulative distribution $F(\eta)$ defined on $[\eta_0, \eta_3]$ in his calculation.

We include side payment as the manufacturer's decision variable to make the contract more flexible and to achieve the supply chain coordination (Corbett et al. 2004). Thus, the profit for the manufacturer is $\pi_M^A = wd_2 + p_1d_1 - L$ and for the retailer is $\pi_R^A = (p_2 - w - c_v)d_2 + L \geq \pi_{\overline{R}}$. This side payment is paid from the manufacturer to the retailer. Then, the retailer's revenue would increase by this amount to cover her fixed cost, while the manufacturer's revenue would reduce by the same amount.

The manufacturer offers a contract to the retailer which is a menu of $\{p_1(\eta), w(\eta), L(\eta)\}$ meaning that it offers a number of alternative values for this tuple. This follows from the Revelation Principle of Fudenberg and Tirole (1991). The retailer has a choice of not accepting the contracts if none of the alternatives are attractive enough to her. Or she may select one alternative from the menu and decides to accept that. The manufacturer maximizes his profit function over the range of η subject to the incentive compatibility (IC) constraint and the individual rationality (IR) constraint that the retailer will at least recover her own reservation profit. This is given in the following formulation.

$$\operatorname*{Max}_{p_1,w,L,N} \int_{\eta_0}^{N} (wd_2 + p_1d_1 - L)f(\eta)\,\mathrm{d}\eta \tag{5.10}$$

subject to

$$\text{Incentive Compatibility constraint or IC: } L = \left(\frac{1}{4\eta} + \frac{a + rp_1 - w}{2}\right)(w - rp) \tag{5.11}$$

$$\text{Individual Rationality constraint or IR: } \pi_R = (p_2 - w - c_v)d_2 + L \geq \pi_{\overline{R}} \tag{5.12}$$

The above formulation follows the standard optimal control formulation with a salvage value Φ. For more details, please refer to Fudenberg and Tirole (1991), Kamien and Schwartz (1981, p. 148), and Mukhopadhyay et al. (2008). The first term in Eq. (5.10) gives the expected value of the manufacturer's profit over the range of η, from the lowest possible value η_0 to the cutoff value N. Equation (5.11) is the IC constraint, forcing the retailer to truly announce her true value η. Equation (5.12) is the IR constraint that make sure the retailer at least earns her reservation profit (or higher profit).

It is not possible to get a close solution for the general asymmetric information, so we give the optimal contract for a special case at the cutoff point when the manufacturer's profit will hit his reservation profit $\pi_{\overline{M}}$.

Proposition 3

(a) *The optimal contract under asymmetric information is given by:*

$$p_1^A = \frac{a}{2(1-r)}, \quad w^A = \frac{F}{2f\eta^2} - \frac{ra}{2(1-r)},$$
$$p_2^A = \frac{a}{2(1-r)} + \frac{3}{4\eta} + \frac{F}{4f\eta^2}, \quad v^A = \frac{1}{\eta}$$
$$\frac{\partial L^A}{\partial \eta} = \left(\frac{a}{2} - \frac{F}{4f\eta^2} + \frac{1}{4\eta}\right)^2 \frac{\partial w^A}{\partial \eta},$$

$\eta_1 = N^A, \eta_1$ satisfies $\frac{\partial \pi_R^A}{\partial \eta} = 0$, η_0 satisfies $\left(p_2^A - w^A - c_v^A\right)d_2 + L^A = \pi_{\overline{R}}$,
$L(N)^A$ satisfies $-\frac{F^2}{8N^4f^2} + \frac{aF}{4N^2f} + \frac{F}{8N^3f} + \frac{a^2(1+r)}{4(1-r)} - L(N)^A = \pi_{\overline{M}}$

(b) The optimal profits of the retailer and the manufacturer under asymmetric information are given by:

$$\pi_R^A = \left(p_2^A - w^A - c_v^A\right)d_2^A + L^A = \left(\frac{a}{2} + \frac{1}{4\eta} - \frac{F}{4f\eta^2}\right)^2 + L^A$$
$$\pi_M^A = p_1^A d_1^A + w_1^A d_2^A - L^A = -\frac{F^2}{8\eta^4 f^2} + \frac{aF}{4\eta^2 f} + \frac{F}{8\eta^3 f} + \frac{a^2(1+r)}{4(1-r)} - L^A$$

In the next subsection, we will compare the optimal contracts in three cases.

5.4 Analysis of the Contracts

We will use both the mathematical analysis and numerical experiment approaches to compare three contracts.

5.4.1 *Mathematical Analysis*

In this section, we compare the manufacturer's, retailer's, and the whole supply chain's profits and contracts under three cases, the channel integration (*I*), full information (*F*), and the asymmetric information (*A*) case. For the asymmetric information (*A*) case, we assume a uniform distribution for η with $f(\eta) = \frac{1}{\eta_3 - \eta_0}$ and $F(\eta) = \frac{\eta - \eta_0}{\eta_3 - \eta_0}$ over the interval $[\eta_0, \ \eta_3]$. η is within the range of $[\eta_0, \ \eta_3]$, with a low bound of η_0 and a high bound of η_3. We summarize our results in Table 5.2.

Proposition 4

(a) *The retailer's optimal value-added amount remains same under three cases,* $v^I = v^F = v^A = \frac{1}{\eta}$.

(b) *The manufacturer sets same direct channel price* p_1 *under I, F, and A, i.e.,* $p_1^I = p_1^F = p_1^A = \frac{a}{2(1-r)}$.

From Proposition 4(a), we see that the optimal level of value-added service provided at the retail store becomes unaffected in three cases. It is also seen that the optimum value of the value added depends only on one parameter, namely η. Therefore, if the retailer can use operational means to increase her efficiency (smaller η) in offering value-added service, the optimal value-added service will be higher $\frac{1}{\eta}$. Many different ways can be used to improve service efficiency, including employee training, adopting new technology, and others. The retail store can offer an employee training program to help sales associates to better understand products and customer needs. Adopting a customer relationship management (CRM) system or some new technologies at retail store can help improve customer's shopping experience. Zhu et al. (2012) reviewed several radio frequency identification (RFID) applications at retail store. For example, Piramuthu (2007) reports RFID application at man's fitting room at a German department store and the Galeria Kaufhof in Essen, part of the Metro retailing group. When a man enters a fitting room to try a suit, a "smart mirror" will tell you what kind of shirt or tie you need to buy with it. An RFID reader on a "smart mirror" in the fitting room determines which clothing has been brought into the room from the RFID tag attached to the apparel and then displays complementary clothing choices or accessories. The system is used in combination with "smart shelves," which can read what merchandise is currently in stock, so that customers can be shown choices in sizes that are available and in various styles and colors. So men buying clothes in this store

Table 5.2 The optimal contracts under three cases

	Channel integration (*I*)	Full information (*F*)	Asymmetric information (*A*)
Manufacturer wholesale price w	NA	$w^F = \frac{a}{2(1-r)} + \frac{1}{4\eta}$	$w^A = \frac{\eta-\eta_0}{2\eta^2} - \frac{ra}{2(1-r)}$
Online direct channel price p_1	$p_1^I = \frac{a}{2(1-r)}$	$p_1^F = \frac{a}{2(1-r)}$	$p_1^A = \frac{a}{2(1-r)}$
Retail channel price p_2	$p_2^I = \frac{a}{2(1-r)} + \frac{3}{4\eta}$	$p_2^F = \frac{(3-r)a}{4(1-r)} + \frac{7}{8\eta}$	$p_2^A = \frac{a}{2(1-r)} + \frac{3}{4\eta} + \frac{\eta-\eta_0}{4\eta^2}$
Value added at retail channel v	$v^I = \frac{1}{\eta}$	$v^F = \frac{1}{\eta}$	$v^A = \frac{1}{\eta}$
Manufacturer's profit		$\pi_M^F = \frac{a}{8\eta} + \frac{(3+r)a^2}{8(1-r)} + \frac{1}{32\eta^2}$	$\pi_M^A = -\frac{\eta_0}{8\eta^3} + \frac{\eta_0^2}{8\eta^4} + \frac{a}{4\eta} - \frac{a\eta_0}{4\eta^2} + \frac{a^2(1+r)}{4(1-r)} - + L^A$
Retailer's profit		$\pi_R^F = \frac{a}{16\eta} + \frac{a^2}{16} + \frac{1}{64\eta^2}$	$\pi_R^A = \frac{a^2}{4} + \frac{\eta_0 a}{4\eta^2} + \frac{\eta_0^2}{16\eta^4} + L^A$
Total profit	$\pi^I = \frac{a}{4\eta} + \frac{a^2}{2(1-r)} + \frac{1}{16\eta^2}$	$\pi^F = \frac{3a}{16\eta} + \frac{(7+r)a^2}{16(1-r)} + \frac{3}{64\eta^2}$	$\pi^A = \frac{a}{4\eta} + \frac{a^2}{2(1-r)} + \frac{3\eta_0^2}{16\eta^4} - \frac{\eta_0}{8\eta^3}$

will get automatic suggestions. The retail store needs to optimize customer relationships at every touch point and every possible way. Improved efficiency will start a ripple effect by allowing the retailer charge a higher retail price (in Fig. 5.3).

From Proposition 4(a), it is interesting to see that the manufacturer does not change his online direct sale price under three different cases. Also, this price is independent of η.

Proposition 5 *Comparing channel integration (I) and asymmetric information (A) case,*

(a) *The total channel profit is higher in (I) than in (A),* $\pi^I > \pi^A$,
(b) *The retail store will set a higher retail price in (A) than in (I),* $p_2^A > p_2^I$.

From Proposition 5(a), we see that the total profit realized under channel integration is always higher than the sum of the retailer's and the manufacturer's profits without such integration. This is generally the same result found in most supply chain literature. From Proposition 5(b), we can see that the double marginalization occurs in the asymmetric information case. This, in turn, reduces the whole supply chain efficiency. Basically, information asymmetry is inefficient for the supply chain as a whole.

Proposition 6 *Comparing the channel integration (I) and full information (F) cases,*

(a) *The total channel profit is higher in (I) than in (F),* $\pi^I > \pi^F$,
(b) *The retail store will set a higher retail price in (F) than in (I),* $p_2^F > p_2^I$.

Combining Proposition 5(a) and Proposition 6(a), we can state that the total channel profit under (*I*) is always higher than that in any other cases, (*A*) or (*F*). Channel integration (*I*) is the first and best case in terms of total supply chain's profit. From Proposition 6(b), it is not surprise to see that the retail store sets a higher retail price in (*F*) than that in (*I*) due to the double marginalization.

5.4.2 Numerical Experiments

In this section, we take the analysis further using extensive numerical experimentation and gain more insights. We did this experiment for two main reasons. One is to verify some of the analytical findings. The other is to gain more insights into the optimal policies especially when the exogenous parameters vary and thereby get some useful managerial guidelines. The numerical values used in this experiment are given in the next table. Note that all these numbers are set in a way which satisfies the model assumptions and they are representative of the general situation considered in this paper.

a	r	$\pi_{\overline{M}}$	$\pi_{\overline{R}}$	η_0	η_3
40	0.500	1100	600	0.03	0.07

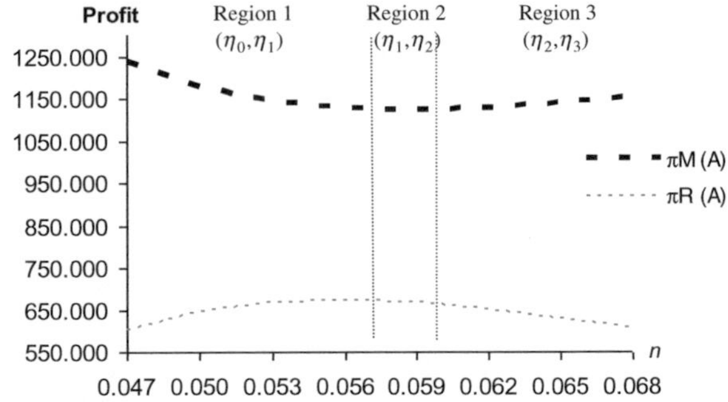

Fig. 5.2 The manufacturer's and retailer's profits under (*A*) for various η

5.4.2.1 Analysis of the Manufacturer's and Retailer's Profits Under (*A*) with Various η

We plot the manufacturer's and retailer's profit under various η as shown in Fig. 5.2. We find out that both of the manufacturer's and the retailer's profits are non-monotone function with η. We can break the range of η, $[\eta_0, \eta_3]$, into three regions. In each of these regions, the profits show monotone trend. For the detailed discussion of how to divide three regions, please refer to Mukhopadhyay et al. (2008).

Region 1: In this region, in the interval of (η_0, η_1), the manufacturer's profit decreases and the retailer's profit increases with η. We get $\eta_0 = 0.03$ and $\eta_1 = 1.91\ \eta_0 = 0.0573$.

Region 2: In the interval of (η_1, η_2), both of the manufacturer's and the retailer's profits decrease with η. $\eta_2 = 2.01\ \eta_0 = 0.0603$.

Region 3: The third interval is given by (η_2, η_3), where $\eta_3 = 0.07$. In this region, the manufacturer's profit increases with η and the retailer's profit decreases with η.

In Region 1, η is relatively small. Please recall that the optimal level of value added = $1/\eta$. So in this region, the level of value added is relatively high. The retail store can provide lot of value-added activities. This phenomenon can be seen in the high-end fashion and luxury brand. For example, customers are more likely to visit a physical store to buy a Louis Vuitton bag or garment. Costumers might need to see and feel the product that he/she is considering before making a purchase decision. Costumers might want to talk to a sales associate in person. The sale associates are well trained and are prepared to give quality customer services. They can give customers some personalized product recommendations based on the face-to-face communication with the customers. The sales associates can also collect customers' feedback to improve the product and build the brand. From Fig. 5.2, we see that the higher the value the retailer adds to the base product, the higher the profit the manufacturer earns. So the manufacturer should encourage the retail store to add more value-added services at the physical store. We also see that the retailer's profit increases with η. The retailer prefers to add small amount of value (with a large η) to avoid heavy cost burden from offering value-added service.

Another extreme case is shown in Region 3. In the region, η is relatively high or the value added from the retail store is relatively low. This phenomenon can be seen for the fast fashion brand, like Forever 21 and H&M. Customers who shop fast fashion clothes focus on the product itself, like the style, and sensitive to the sale price. Customers will prefer to shop online due to the lower online price ($p_1 < p_2$, or the online price < retail channel price). Many customers make a purchase with a few clicks and do not have any questions about the product, or are content with the online customer support. Figure 5.2 shows that the manufacturer's profit increases with η and the retailer's profit decreases in η. We can see that when the value-added level is low, the retail store will suffer a lot from online sale competition, but the manufacturer will benefit from this competition receiving revenue from direct sale channel and wholesale to the retailer. Our study suggests that the fast fashion brand should open a direct channel and let two channels compete each other.

Region 2 is in the middle of two extreme cases. We see that both the manufacturer's and the retailer's profits decrease in η and increase in value added. The high degree of value added at retail store and convenient online channel satisfies customers different shopping preference and reduces channel conflict. Both the manufacturer and retailer would prefer a higher value-added service at retail store and earn higher profits.

We can also apply the above "three regions" discussion to different product lines offered by a same company. For example, Gap Inc. has three product lines, including Old Navy, Gap, and Banana Republic. Old Navy sells to value-seeking families; Gap attracts trend-conscious shoppers; Banana Republic offers the company's higher-end label clothing. For Banana Republic, filling in Region 1, the company can consider to improve the value-added service or customer service at the store. For Old Navy, filling in Region 3, the company can more focus on online channel sales. Gap is filling in Region 2.

5.4.2.2 Analysis of the Retail Channel Price with Various η

From Proposition 4(b), we find that the online channel price is always same for case (*I*), (*F*), and (*A*). Figure 5.3 shows the retail channel price under (*I*), (*F*), and (*A*) is decreasing with η. The lower is the level of value-added service (higher η), the lower is the retail price at the retail channel. The findings from Fig. 5.3 are consistent with our mathematic analysis in Proposition 5(b), $p_2^A > p_2^I$, and Proposition 6 (b), $p_2^F > p_2^I$. We see that the retail channel price under (*I*) case is the lowest among three cases. As we discussed earlier in Proposition 5 and Proposition 6, the channel integration is always the first and best case. The customers get the lowest retail channel price, and manufacturer gets the highest profit under channel integration. So when possible, the manufacturer should treat the retail channel and online channel as a whole and maximize the total channel profit.

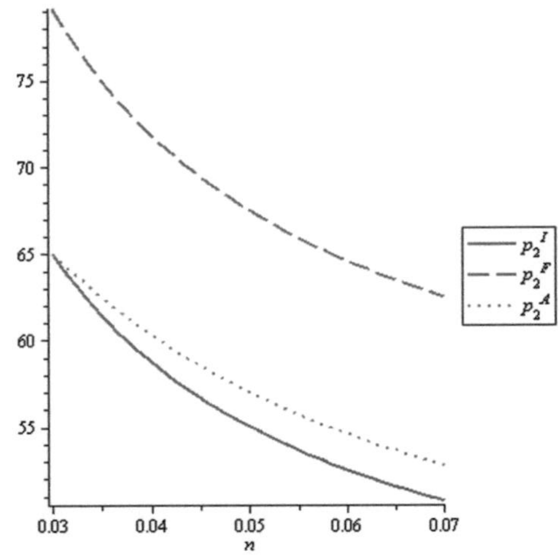

Fig. 5.3 The retail channel price under (*I*), (*F*), and (*A*) for various η

5.4.2.3 Analysis of the Retail Channel Price with Various *r*

From Fig. 5.4, it is interesting to see that the retail channel prices are increasing with channel migration factor *r* for all three cases of (*I*), (*A*), and (*F*). Higher value of *r* means higher channel migration. The ways to increase *r* include let customer better aware of attractive online channel price and value-added service at retail store.

5.4.2.4 Analysis of the Value of Information

In Fig. 5.5, we plot the manufacturer's and retailer's profits under (*F*) and (*A*) with various η. We see the manufacturer earns a higher profit under (*F*) than under (*A*),

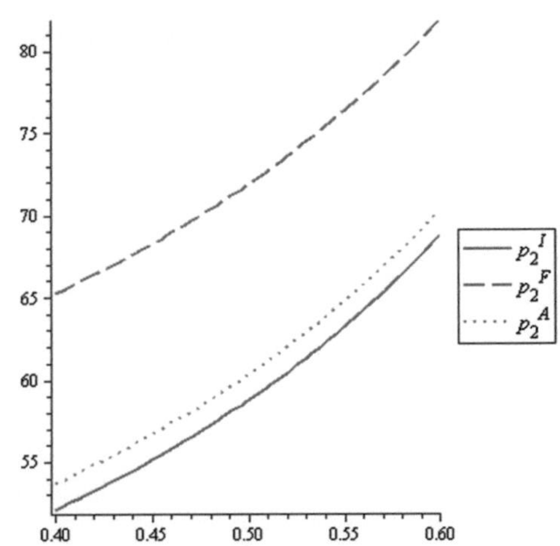

Fig. 5.4 The retail channel price under (*I*), (*F*), and (*A*) for various *r*

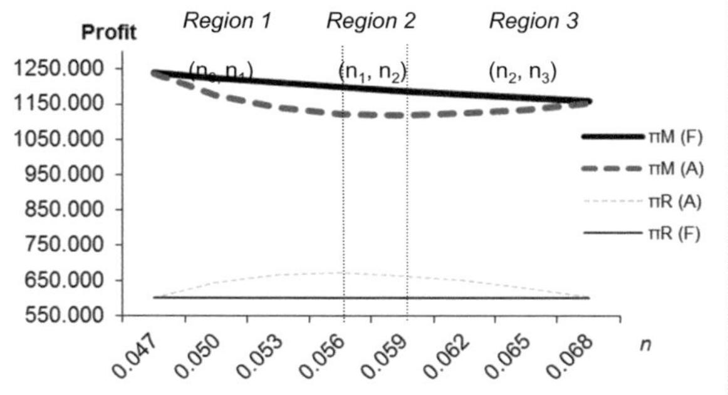

Fig. 5.5 The manufacturer's and retailer's profits under (*A*) and (*F*) for various η

but the retailer earns a higher profit under (*A*) than under (*F*). It is not surprise because the retailer holds her private information as a secret under (*A*) and the manufacturer has information disadvantage under (*A*). The profit difference between (*A*) and (*F*) is the value of information. The gap of $\pi_M^I - \pi_M^A$ is the value of information for the manufacturer and $\pi_R^A - \pi_R^I$ is the value of information for the retailer.

We also carried out sensitivity analyses on the effect on profit of other parameters such as base demand a and the migration parameter r. Because of length consideration, we do not report the entire findings here. In short, we find that, with the base demand a increasing, the value of information (given by the difference between the profits in A and I cases) increases. The managerial guideline here is that the retailer should try to increase the base demand by means of, say, advertising or offering better return policy. We also found that when the migration parameter r increases, the value of information increases. Again, the guideline for the retailer is that she should use marketing means to influence r.

We have summarized the findings in this section in the following table.

Figure	Summary of findings
Figure 5.2	As indicated in Region 1, customers prefer to visit a physical store to shop the high-end fashion and luxury brand products. The manufacturer should encourage the retail store to add more value-added services at the physical store. But the retailer prefers to add small amount of value (with a large η) to avoid heavy cost burden from offering value-added service As indicated in Region 3, customers prefer to shop online for fast fashion clothes. The fast fashion brand should open a direct channel in addition to the retail store and let two channels compete each other As indicated in Region 2, both of the manufacturer's and the retailer's profits decrease in η and increase in value added. Both of them prefer a higher value-added service at retail store and earn higher profits Company can consider differential products or promotion to reduce the channel conflicts
Figure 5.3	The retail channel price under (*I*) case is the lowest among three cases of (*I*), (*A*), and (*F*)

(continued)

(continued)

Figure	Summary of findings
	The channel integration will always the first and best case
Figure 5.4	The retail channel prices are increasing with channel migration factor *r* for all three cases of (*I*), (*A*), and (*F*)
Figure 5.5	The manufacturer earns a higher profit under (*F*) than under (*A*) but the retailer earns a higher profit under (*A*) than under (*F*)

5.5 Conclusions

In this paper, we study the fashion clothing sale through a mixed channel where the manufacturer has a direct online channel in addition to the traditional retail store channel. The retail store offers some value-added services to the end customers. We use a game theoretic formulation for this channel sale problem under three different scenarios, including channel integration (*I*), under full information (*F*), and under asymmetric information (*A*). We obtained closed-form contracts for three cases and compare the results by using mathematical analysis and numerical experiment approaches.

We find that the quantum of value added does not change under any scenario and is only dependent on the retailer's cost structure η. Improved efficiency (small η) will start a ripple effect by allowing the retailer to charge higher retail price p_2 and offer high level of value-added service at retail store. The retail channel price under channel integration (*I*) p_2^I gives the lowest sale price among three cases. It is interesting to see that the online price p_1 also does not change under three different cases. We discuss the sale channel structure based on various brand strategies, such as high-end fashion, fast fashion, and brand between these two extremes and give suggestions to improve the whole channel operations. Information asymmetry imposes inefficiency to the manufacturer and to the supply chain as a whole. The channel integration (*I*) is always the first and best case.

We hope this research can help practitioners in fashion industry to improve the sale channel design and whole channel efficiency. Our model can be extended in many different directions. For example, we can let the manufacturer to offer the value-added service through online channel, instead of at the retail store discussed in this paper. We can discuss how the return policy affects the channel design by including a return rate factor in the demand function. Advertising strategy for mixed channel is also an interesting topic. We can also discuss the manufacturer's and retailer's profit and price policy for the omni-channel that can include retail stores, online stores, mobile stores, telephone sales, TV sales, and other channels. The practitioners in fashion industry might be interested in learning how to use the omni-channel to launch new product and sell season ending excess inventory.

Appendix

Proof of Proposition 1(a)

$$\begin{aligned}\pi^I &= p_1 d_1 + (p_2 - c_v) d_2 \\ &= p_1(a - p_1 + r(p_2 - v)) + \left(p_2 - \frac{\eta v^2}{2}\right)(a - p_2 + v + p_1 r)\end{aligned}$$

Then, we take first-order condition with respect to p_1, p_2, and v and set them equal to zero, respectively. After that, solving these three equations simultaneously, we can get the desired result.

Proof of Proposition 1(b)

$$\begin{aligned}&\pi_R^F = (p_2^* - w - C_v^*) d_2 + L^F = \pi_{\overline{R}} \\ &\Rightarrow L^F = \pi_{\overline{R}} - \left(\tfrac{2a\eta + 1}{4\eta}\right)^2\end{aligned}$$

$$\pi_M = \left\{\begin{array}{ll} \frac{a}{4\eta} + \frac{1}{16\eta^2} + \frac{a^2}{2(1-r)} - \pi_{\overline{R}} & \underline{\eta} \le \eta \le N \quad (a) \\ \pi_{\overline{M}} & N \le \eta \le \overline{\eta} \quad (b) \end{array}\right\}$$

Setting (a) = (b), we get

$$N = \frac{1}{-2a + 2\sqrt{4\pi_{\overline{M}} + 4\pi_{\overline{R}} - a^2 \frac{1+r}{1-r}}} \qquad \text{where } \pi_{\overline{M}} + \pi_{\overline{R}} \ge \frac{a^2(1+r)}{4(1-r)}$$

Due to $N > 0$, we only keep the one with positive value.

Proof of Proposition 3 The Eqs. (5.10), (5.11), and (5.12) can be written as follows:

$$\begin{aligned}&\max \int_{\underline{\eta}}^{N} m(\eta) \mathrm{d}\eta + \Phi(N) \\ &\text{s.t.} \\ &\dot{L}(\eta) = g_1(\eta),\ \dot{w}(\eta) = g_2(\eta),\ \dot{p}_1(\eta) = g_3(\eta)\end{aligned}$$

This is obtained by making the following variable substitution:

$$m: = (p_1 d_1(p_2^r) + wd_2(p_2^r) - L)f$$
$$= \left\{p_1\left[\left(1+\frac{r}{2}\right)a + \left(\frac{r^2}{2}-1\right)p_1 - \frac{r}{4\eta}\right] + w\left(\frac{a}{2}+\frac{1}{4\eta}-\frac{w}{2}\right) + rwp_1 - L\right\}f,$$
$$g_1 = \left(\frac{1}{4\eta}+\frac{a+rp_1-w}{2}\right)(u_1 - ru_2),\ \ g_2 = u_1,\ \ g_3 = u_2,$$
$$\Phi(N) = \pi_{\overline{M}}(1-F).$$

Using the multiplier equations gives following results:

$$\dot{\lambda}_1 = f \text{ and } \lambda_1 = F \tag{5.13}$$

$$\dot{\lambda}_2 = -\left(-w+\frac{a}{2}+\frac{1}{4\eta}+rp_1\right)f + \frac{\lambda_1}{2}(u_1 - ru_2) \tag{5.14}$$

$$\dot{\lambda}_3 = -\left[a\left(1+\frac{r}{2}\right) + 2p_1\left(\frac{r^2}{2}-1\right) - \frac{r}{4\eta} + rw\right]f - \frac{\lambda_1 r(u_1 - ru_2)}{2} \tag{5.15}$$

Using the optimality conditions gives following results:

$$\lambda_1\left(\frac{1}{4\eta}+\frac{a+rp_1-w}{2}\right) + \lambda_2 = 0 \tag{5.16}$$

$$-r\lambda_1\left(\frac{1}{4\eta}+\frac{a+rp_1-w}{2}\right) + \lambda_3 = 0 \tag{5.17}$$

Taking derivative on both sides of (5.16) and using (5.13), we get

$$\dot{\lambda}_2 = -\left(\frac{1}{4\eta}+\frac{a+rp_1-w}{2}\right)f - F\left(-\frac{1}{4\eta^2}+\frac{r}{2}u_2-\frac{u_1}{2}\right) \tag{5.18}$$

Solving (5.18) with (5.14), we get

$$\frac{F}{2\eta^2} = fw - frp_1 \tag{5.19}$$

Taking derivative on both sides of (5.17) and using (5.13), we get

$$\dot{\lambda}_3 = rf\left(\frac{1}{4\eta}+\frac{a+rp_1-w}{2}\right) + rF\left(-\frac{1}{4\eta^2}+\frac{r}{2}u_2-\frac{u_1}{2}\right) \tag{5.20}$$

Solving (5.20) with (5.15), we get

$$\frac{rF}{4\eta^2} = f\left[a + ar + \left(\frac{3r^2}{2} - 2\right)p_1 + \frac{rw}{2}\right] \tag{5.21}$$

Solving (5.19) and (5.21) together, we get desired result

$$p_1^A = \frac{a}{2(1-r)}, w^A = \frac{F}{2f\eta^2} - \frac{ra}{2(1-r)}$$

and

$$\dot{L}(\eta) = g_1(\eta) = \left(\frac{1}{4\eta} + \frac{a + rp_1 - w}{2}\right)(u_1 - ru_2) = \left(\frac{1}{4\eta} + \frac{a + rp_1 - w}{2}\right)\dot{w}.$$

Using the transversality conditions if N is free

$$m(N) + \lambda_1(N)g_1(N) + \lambda_2(N)g_2(N) + \lambda_3(N)g_3(N) + \Phi_N = 0 \,\text{at}\, N$$

we get the following results: $(p_1d_1 + wd_2 - L - \pi_{\overline{M}})f = 0$. Because $f \neq 0$, $p_1d_1 + wd_2 - L - \pi_{\overline{M}}$ must equals to 0. The manufacturer can make $p_1d_1 + wd_2 - L - \pi_{\overline{M}} \geq 0$ binding at η_1, $\eta_1 = N$. Then, substituting p_1 and w with $p_1^A(N)$ and $w^A(N)$, we get that $L(N)^A$ satisfies $-\frac{F^2}{8N^4f^2} + \frac{aF}{4N^2f} + \frac{F}{8N^3f} + \frac{a^2(1+r)}{4(1-r)} - L(N)^A = \pi_{\overline{M}}$.

η_0 can be solved by let $\left(p_2^A - w^A - c_v^A\right)d_2 + L^A \geq \pi_{\overline{R}}$ binding at η_0.

Proof of Proposition 5

(a) $\pi^I - \pi^A = \frac{(\eta-\eta_0)(\eta+3\eta_0)}{16\eta^4}$. Because $\eta > \eta_0$, $\eta > 0$, and $\eta_0 > 0$, $\pi^I - \pi^A > 0$.
(b) Because the proof is straightforward, we omit the proof here.

Proof of Proposition 6

(a) $\pi^I - \pi^F = \frac{(1+2a\eta)^2}{64\eta^4} > 0$
(b) $p_2^F - p_2^I = \frac{1+2a\eta}{8\eta} > 0$

References

Aviv Y (2002) Gaining benefits from joint forecasting and replenishment processes: the case of auto-correlated demand. Manuf Serv Oper Manage 4(1):55–74

Balasubramanian S (1998) Mail versus mall: a strategic analysis of competition between direct marketers and conventional retailers. Mark Sci 17:181–195

Beck N, Rygl D (2015) Categorization of multiple channel retailing in multi-, cross-, and Omni-channel retailing for retailers and retailing. J Retai Consum Serv 27:170–178

Chiang W, Chhajed D, Hess JD (2003) Direct marketing, indirect profits: a strategic analysis of dual-channel supply chain design. Manage Sci 49(1):1–20

Cohen M, Agrawal N, Agrawal V, Raman A (1995) Analysis of distribution strategies in the industrial paper and plastic industries. Oper Res 43(1):6–18
Collinger T (1998) Lines separating sales channels blur: manufacturers, direct sellers, retailers invade each other's turf. Advertising age. Last accessed in 30 July 2015 at http://adage.com/article/news/lines-separating-sales-channels-blur-manufacturers-direct-sellers-retailers-invade-s-turf/66410/
Corbett C, de Groote X (2000) A supplier's optimal quantity discount policy under asymmetric information. Manage Sci 46(3):444–450
Corbett C, Deming Z, Tang CS (2004) Designing supply contracts: contract type and information asymmetry. Manage Sci 50(4):550–559
Desiraju R, Moorthy S (1997) Managing a distribution channel under asymmetric information with performance requirement. Manage Sci 43(12):1628–1644
Fay CJ (1999) The direct approach. Best's Rev 99(11):91–93
Frazier GL (1999) Organizing and managing channels of distribution. J Acad Mark Sci 27(2):226–241
Fudenberg D, Tirole J (1991) Game theory. The MIT Press
Gan X, Sethi SP, Cakanyildirim M (2003) Contracting under asymmetric information about production cost. Working Paper, University of Texas at Dallas
Gavirneni S (2002) Information flows in capacitated supply chains with fixed ordering costs. Manage Sci 48(5):644–651
Gavirneni S, Kapuscinski R, Tayur S (1999) Value of information in capacitated supply chains. Manage Sci 45(1):16–24
Ha AY (2001) Supplier-buyer contracting: asymmetric cost information and cutoff level policy for buyer participation. Naval Res Logistics 48:41–64
Hübner A, Holzapfel A, Kuhn H (2015) Operations management in multi-channel retailing: an exploratory study. Oper Manage Res, 1–17
Kamien MI, Schwartz N (1981) Dynamic optimization: the calculus of variations and optimal control in economics and management. North Holland
Kozlenkova IV, Hult GTM, Lund DJ, Mena JA, Kekec P (2015) The role of marketing channels in supply chain management. J Retai 90(4):586–609
Lee H, So KC, Tang C (2000) The value of information sharing in a two-level supply chain. Manage Sci 46:626–643
Levary R, Mathieu RG (2000) Hybrid retail: integrating e-commerce and physical stores. Ind Manage 42(5):6–13
Mau D (2014) Why some luxury brands still don't sell online. Last accessed in 30 July 2015 at http://fashionista.com/2014/03/why-some-luxury-brands-still-dont-sell-online
Mukhopadhyay S, Zhu X, Yue X (2008) Optimal contract design for mixed channels under information asymmetry. Prod Oper Manage 17(6):641–650
Piramuthu S (2007) Protocols for RFID tag/reader authentication. Decis Support Syst 43(3):897–914
Tsay AA, Agrawal N (2004) Channel conflict and coordination in the e-commerce age. Prod Oper Manage 13(1):93–110
Tsay AA, Nahmias S, Agrawal N (1999) Modeling supply chain contracts: a review, in quantitative models for supply chain management. In: Tayur S, Ganeshan R, Magazine M (eds) Kluwer Academic Publishers
Verhoef PC, Kannan PK, Inman JJ (2015) From multi-channel retailing to Omni-channel retailing: introduction to the special issue on multi-channel retailing. J Retai 91(2):174–181
Viswanathan S (2000) Competition across channels: do electronic markets complement or cannibalize traditional retailers. In: Proceedings of the twenty first international conference on Information systems. International conference on information systems
Yao D, Liu J (2003) Channel redistribution with direct selling. Eur J Oper Res 144:646–658
Zhu X, Mukhopadhyay S, Kurata H (2012) A review of RFID technology and its managerial applications in different industries. J Eng Technol Manage, 152–167

Chapter 6
Impacts of Vendor-Managed Strategic Partnership on Fashion Supply Chains with Markdown Money Policy

Bin Shen, Rongrong Qian and Vincent Quan

Abstract A vendor-managed inventory (VMI) partnership is a popular approach to promote channel performance. This paper is motivated by real-industrial practices in the fashion industry and explores the issue of how a VMI partnership with markdown money policy (MMP) operates in the fashion supply chain. MMP is also known as a vendor agreement whereby a vendor supports the profitability of their particular brand with a specific retailer. A vendor allowance (VA) in the form of markdown money is issued to the said retailer in a certain period, e.g., every quarter or six-month season. We propose a model in the context of a two-echelon supply chain with one single supplier and one single retailer trading via a VMI partnership with MMP in both decentralized and centralized supply chains. We find that under the VMI mode with an MMP, the supply chain is able to achieve coordination, and the retailer's profit is better off but the supplier suffers. We then conduct a numerical study to further explore the impact of VMI within the supply chain with the MMP. Important insights into industry practices are discussed.

Keywords Vendor-managed inventory · Markdown money policy · Fashion supply chain · Vendor allowance · Gross margin agreement

B. Shen (✉) · R. Qian
Glorious Sun School of Business and Management, Donghua University,
1882 Yanan West Road, Shanghai 210005, China
e-mail: Binshen@dhu.edu.cn

R. Qian
e-mail: qr_1994@163.com

V. Quan
Jay and Patty School of Business and Technology,
Fashion Institute of Technology, State University of New York,
Seventh Avenue at 27 Street, New York, USA
e-mail: vincent_quan@fitnyc.edu

T.-M. Choi (ed.), *Analytical Modeling Research in Fashion Business*,
Springer Series in Fashion Business, DOI 10.1007/978-981-10-1014-9_6

6.1 Introduction

A vendor-managed inventory (VMI) partnership is a program recognized as one of the most successful practices in the fashion industry which enhances cooperation and improves efficiency throughout the supply chain. To improve supply chain efficiency, many fashion suppliers have adopted the VMI partnership model to trade with their retailers (Zhang 2010; Chen et al. 2012). Under a VMI program, the supplier is responsible for monitoring the retailer's inventory levels and makes ordering suggestions on behalf of the retailer. The retailer then reviews all orders and approves or disapproves them. The vendor generates revenue through shipments to the retailer while the retailer generates revenue by selling the products to consumers. The well-known story of a successful VMI partnership is JC Penney (i.e., retailer) and TAL (vendor) (Wong et al. 2009). In addition, the Chinese fashion brand Lady Forever (vendor) also adopted a VMI program to trade with its retail franchisees (Zhang 2010).

The VMI is usually applied with the supply chain contracts that include limits on a retailer's upper inventory level such that the supplier has to pay a penalty for exceeding agreed-upon limits (Shah and Goh 2006; Chen et al. 2006). Similar to this, many fashion suppliers provide incentive contracts such as a markdown money policy (MMP) to encourage retailers to join in the VMI partnership. For example, Lady Forever adopts the MMP in their supply chains (Zhang 2010). The MMP has the effect of motivating the retailer to order more merchandise which mitigates the double marginalization effect (Wang and Webster 2009). The supplier (i.e., vendor) will provide the supportive money for markdowns at the end of selling season avoiding returns of the physical product by the retailer. It should be noted that there are cases where an MMP includes both VA money and retailer returns authorized by the supplier. Given this supportive money and partnership, the retailer is willing to cooperate with her upstream supplier and provide access to inventory control (Chow 2015). Meanwhile, the supplier can prevent the retailers from discounting the overstocking fashion goods by providing an MMP and authorizing returns to avoid damaging its own brand image at the hands of the retailer. However, it is unknown that whether the MMP can help supply chain achieve channel coordination under the VMI partnership.

In this paper, we are motivated by the practices of VMI adoption in the fashion supply chain and examine how the VMI program affects the supply chain with the MMP. We consider a two-echelon supply chain with the MMP under a VMI partnership. We examine how an MMP affects supply chain performance.

The remainder of this paper is organized as follows: In Sect. 6.2, we review the related literature. Section 6.3 describes our model from both decentralized and centralized scenarios, respectively, and Sect. 6.4 conducts a numerical study. Section 6.5 concludes the paper. All of the technical proofs are relegated to the Appendix.

6.2 Literature Review

The VMI partnership is a popular practice in the fashion supply chain. The previous literature has discussed the benefits of VMI implementation (for more details, please refer to Choi et al. 2004). Disney and Towill (2003) found that the adoption of VMI can offer a significant opportunity to reduce the bullwhip effect in real-world supply chains. Cetinakaya and Lee (2000) presented an analytical model to compute the optimum replenishment quantity and shipment scheduling for the VMI system with a shipment-release policy. Dong and Xu (2002) compared the performance of VMI partnership in the short-term with that in the long-term and indicated that VMI could increase a supplier's profit more so in the long-term than in the short-term. Tyan and Wee (2003) examined the practical implementation of the system in the Taiwanese grocery industry. They found that a VMI partnership reduces costs and improves service levels while creating business opportunities for both parties in the supply chain. Yao et al. (2007) explored how the VMI agreement generates benefits in the form of inventory cost reductions from integration. They showed that these benefits cannot be shared between both retailers and suppliers proportionally. Mishra and Raghunathan (2004) indicated VMI intensifies brand competition between manufacturers. However, there is a limited amount of literature addressing the issue how VMI partnerships and contracts affect supply chain performance jointly.

The previous literature has examined some supply chain contracts under the VMI mode. Bernstein et al. (2006) proved that perfect coordination can be achieved within a simple wholesale pricing scheme through a VMI partnership. Nagarajan and Rajagopalan (2008) examined the contracting terms under VMI systems using holding cost subsidies. They evaluated this contract in three inventory schemes and showed when such contracts may improve supply chain performance. Their model adopts the holding cost subsidy-type contracts but does not take markdown money into account. Wong et al. (2009) studied a two-echelon supply chain with a single supplier serving multiple retailers and showed how a sales rebate contract helps achieve channel coordination. They argue that the VMI partnership facilitates the application of the sales rebate contract.

In this paper, we examine the MMP, which is commonly adopted in the fashion industry (Choi 2013a, b). Note that MMP has recently been examined in different contexts. For instance, Ning et al. (2011) investigate the MMP in a supply chain with the presence of an online market. They find that the MMP can help coordinate the corresponding supply chain system. Shen et al. (2013) study the impact of supplier's risk-aversion in fashion supply chain coordination with MMP. They find that risk attitude significantly affects the optimal contract setting in the supply chain. According to previous literature in the VMI partnership with contracts, the practice of how the MMP–VMI joint program affects the channel performance has rarely been documented. Its application in the fashion industry is also unknown. This paper hence helps to fill this important gap. A comparison study is shown in Table 6.1.

Table 6.1 Literature review: a comparison

Article	VMI	MMP	Fashion industry
Cetinakaya and Lee (2000)	√		
Dong and Xu (2002)	√		
Disney and Towill (2003)	√		
Tyan and Wee (2003)	√		
Mishra and Raghunathan (2004)	√		
Bernstein et al. (2006)	√		
Yao et al. (2007)	√		
Nagarajan and Rajagopalan (2008)	√		
Wong et al. (2009)	√		
Ning et al. (2011)		√	
Shen et al. (2013)		√	√
Choi (2013a)			√
Choi (2013b)			√
Our paper	√	√	√

6.3 The Model

In this section, we consider a two-echelon supply chain in which there is one single supplier (she) and one single retailer (he) who are trading via a VMI program with an MMP. Under the VMI program, the supplier is responsible for deciding replenishment quantity. We consider product cost is c per unit and the product is sold to the retailer at a wholesale price w per unit. The retailing price is r per unit and the salvage value is v per unit.

Under the MMP, we consider that all overstocked products will be paid at b per unit as supportive money for markdowns at the end of selling season. Compared with the pure wholesale price contract, the supplier shares a portion of the cost associated with unsold products under an MMP. To ensure incentive mechanisms work in the supply chain, we assume that $0 \leq v < c < w < r$, $0 \leq b \leq w - v$.

Following the practices in the VMI fashion supply chain, the supplier–retailer transaction can be treated as a Stackelberg game in which the retailer is the leader who first decides the contract parameter b in the MMP, and the supplier is the follower who then decides the order quantity q. Therefore, following the sequence of the game above, we can use backward induction to identify the optimal solution. We consider that market demand is random. To facilitate the model, we denote it as a random variable x with a probability density function $f(x)$ and a corresponding cumulative distribution function $F(x)$ that has a unique inverse function $F^{-1}(x)$.

To facilitate the analysis, we consider the retail price r and wholesale price w are exogenous but the markdown price b is endogenous in the VMI fashion supply chain. This assumption is supported by many apparel industrial applications. As in the most case, the wholesale price and retail price of fashion products are determined before the MMP is offered, but the markdown price is decided according to the market conditions. We discuss two scenarios of supply chain structure in the VMI partnership: a decentralized supply chain and a centralized supply chain. Let *EP* represents the expected profit. We use $\mathrm{EP_{SC}}$, $\mathrm{EP_S}$, and $\mathrm{EP_R}$ to represent the expected profit of supply chain, supplier, and retailer, respectively. Moreover, the asterisk is denoted as the optimal level of the parameter in Tables 6.3 and 6.4.

6.3.1 Decentralized Supply Chain

In this section, we first consider a decentralized supply chain scenario. As the wholesale price w per unit is given, the retailer (i.e., Stackelberg leader) offers the markdown price b per unit to the supplier (i.e., follower), and the supplier reacts by determining quantity to maximize her own expected profit. According to backward induction, we first derive the supplier's expected profit with respect to q. The supplier's expected profit is given as follows:

$$\mathrm{EP_S}(q) = (w - c)q - b\int_0^q F(x)\mathrm{d}x, \tag{6.1}$$

where the first item represents the profit from the supply quantity q, and the second item represents the expected cost, which assesses markdown money b granted from the supplier for each of the expected leftover products. In this case, the supplier's objective is to maximize her expected profit. After taking the first and second derivatives with respect to q, we can know $\mathrm{EP_S}(q)$ is strictly concave in q and obtain the supplier's optimal quantity provided to the retailer as

$$q_{\mathrm{S}^*} = F^{-1}[(w - c)/b]. \tag{6.2}$$

Proposition 1 *Both* $\mathrm{EP_S}(q_{\mathrm{S}^*}(b))$ *and* $q_{\mathrm{S}^*}(b)$ *are decreasing in b,* $\forall 0 \le b \le w - v$.

The result of Proposition 1 shows that the higher markdown money b required by retailer, the less q will be provided by the supplier. And the supplier will get less profit with a lower providing quantity.

By substituting the supplier optimal quantity $q_{\mathrm{S}^*}(b)$ into the retailer's profit function, we can yield the following:

$$\mathrm{EP_R}(q_{\mathrm{S}^*}(b)) = (r - w)q_{\mathrm{S}^*}(b) - (r - b - v)\int_0^{q_{\mathrm{S}^*}(b)} F(x)\mathrm{d}x. \tag{6.3}$$

We derive the supplier expected profit with respect to b and obtain the following proposition.

Proposition 2 *Let* $b_0^* = \max(\arg\{\mathrm{dEP_R}(b, q_{\mathrm{S}^*}(b))/\mathrm{d}b = 0\})$, *(i) if* $w - v \ge b_0^* \ge 0$, *then* $b^* = b_0^*$; *(ii) if* $b_0^* > w - v$, *then* $b_0^* > w - v$.

Proposition 2 implies that it is not wise for the retailer to set the upper bound value $(w - v)$ as the markdown money in all instances since the expected profit of the retailer is not strictly increasing in the markdown money b. If the value of b_0^* exceeds the upper bound, we set the optimal b^* equals $(w - v)$. Under this condition, the retailer is risk free for the leftover products.

6.3.2 *Centralized Supply Chain*

In this section, we consider a centralized supply chain under a VMI program. The expected profit of supply chain is $\mathrm{EP_{SC}} = \mathrm{EP_R} + \mathrm{EP_S} = (r - c)q - (r - v) \int_0^q F(x)\mathrm{d}x$. We derive the expected profit of supply chain with respect to q, and we yield $q_{\mathrm{SC}^*} = F^{-1}[(r - c)/(r - v)]$. To achieve supply chain coordination, the supplier must provide the corresponding quantity to the retailer, i.e., $q_{\mathrm{S}^*} = q_{\mathrm{SC}^*}$ (Cachon 2003). Then, we can know

$$b_{\mathrm{SC}}^* = \frac{(w - c)(r - v)}{(r - c)}. \tag{6.4}$$

When supply chain achieves channel coordination under the VMI program, we can know the following proposition that how MMP affects channel's profit distribution.

Proposition 3 *The value of optimal buyback price b_{SC} can allocate the centralized optimal profit between the supplier and the retailer in the VMI mode.*

Proposition 3 implies that the value of optimal buyback price can allocate the centralized optimal profit between the supplier and the retailer. It is not surprising that our result is similar to the case in the retailer-managed inventory (RMI) mode (please refer to Cachon 2003 for the details of RMI). The quantity of products decided in the centralized way is the same in both the VMI mode and the RMI mode. The supply chain profit would not change from adopting the traditional RMI mode to the VMI one. Since the expected profit of retailer is increasing in b_{SC}, compared with the case without the markdown price, i.e., $b_{\mathrm{SC}} = 0$, the retailer could gain an extra profit from the leftovers. To be specific, we analyze the per-unit loss of the leftovers which are listed in Table 6.2. Moreover, we can see that the MMP does not change the supply chain's profit, but affects the allocation of supplier and the retailer's profit. We find that when $c - w + b \geq 0$, the supplier could earn a

Table 6.2 The unit loss of the leftovers for supply chain, supplier, and retailer

Per leftover cost	Pure wholesale price policy	Wholesale price policy and MMP	Margin
Supply chain	$c - v$	$c - v$	0
Supplier	$c - w$	$c - w + b$	b
Retailer	$w - v$	$w - b - v$	$-b$

profit, and vice versa while without the markdown money, she can earn the profit of $(w - c)$ per unit from the residual inventory.

Proposition 4 *The optimal profit of retailer (supplier) is larger (smaller) in the decentralized supply chain than that in the coordinated one.*

Comparing the optimal expected profit in the decentralized and centralized cases, we find that the optimal profit of retailer (supplier) is larger (smaller) in the decentralized supply chain than that in the centralized one when achieving channel coordination. This result reveals that the retailer and the supplier may not reach an agreement to coordinate the supply chain. To force the retailer to participate in the supply chain under the VMI mode, two possible motivations are stated as follows. First, the supplier who has a strong bargaining power in supply chain would force the retailer to join. Second, by joining the program, the retailer may receive more responsive services from the supplier in inventory management and make more efforts on selling other products.

6.4 Numerical Study

We now use two examples to illustrate the optimal markdown price b under the decentralized and centralized supply chains, respectively. We set $r = 200$, $c = 50$, and $v = 5$. The market demand for the product is assumed to be normally distributed with mean 100 and standard deviation 30.

In example 1, we set $w = 100$ in the general condition, and the optimal b can be calculated via $b^* = \arg\{b|\mathrm{dEP_R}(b, q_{S^*}(b))/\mathrm{d}b = 0\}$. Then, we have $q^* = 110$, and the optimal $\mathrm{EP_R} = 8961.02$. In example 2, we set $w = 120$ to satisfy a sufficient condition that $\mathrm{dEP_R}(b, q_{S^*}(b))/\mathrm{d}b|_{b=w-v} \geq 0$, i.e., $\mathrm{EP}'_{\mathrm{R}}(w - v) > 0$ for the retailer to set $b^* = w - v$. In this case, we get $b^* = 126.80$ without constraints on the region of b and the corresponding $\mathrm{EP_R} = 7360.24$. When the markdown money is no larger than $w - v$, thus in this case $b^* = 115.00$ and the corresponding $\mathrm{EP_R} = 7337.70$. Note the example 2 (i.e., MMP 2-1, 2-2), in which we calculate two values of b in the same condition and is analogous to the Proposition 1. Chart 6.1 shows the expected profit of the retailer with markdown money b and indicates the optimal b in different cases with different wholesale prices w.

As the numerical examples are represented in Table 6.3 and Chart 6.1, it is obvious that the retailer can obtain the higher profit with the MMP. For the supplier, the optimal quantity is lower and the expected profit decreases when the markdown price is offered. This implies that MMP is a profitable contract for the retailer, while it may not be wise for the supplier to provide markdown money.

Table 6.4 shows that how the MMP works in the VMI fashion supply chain. Chart 6.2 indicates the expected profit functions of the retailer, supplier, and supply chain with q. We also set $r = 200$, $c = 50$, $v = 5$, and $w = 100$. The market demand for the product is assumed to be normally distributed with mean 100 and

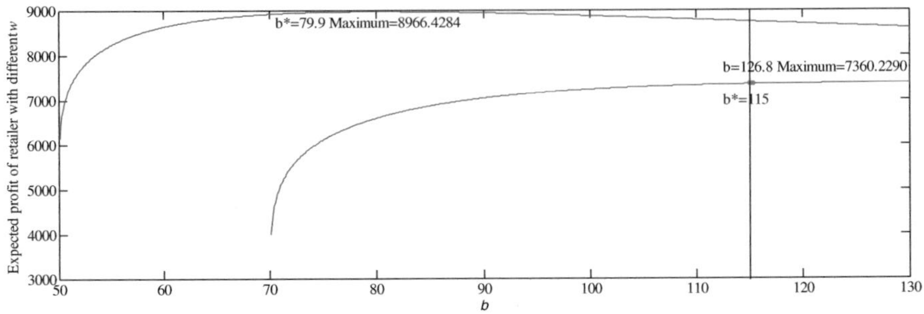

Chart 6.1 Optimal markdown money b in two cases with different w in the decentralized supply chain

Table 6.3 Numerical examples under the decentralized SC

MMP	w	b^*	q^*	EP_R	EP_S	$EP'_R(w-v)$	EP_{SC}
1	100	79.95	110	8961.02	4091.02	−6.70	13,052.03
2-1	120	115.00	108	7337.70	5675.44	3.61	13,013.14
2-2	120	126.80	104	7360.24	5489.85	3.61	12,850.09

Table 6.4 Changes of q and EP with different b

MMP case	b	q	EP_R	EP_S	EP_{SC}
1	60	129	8625.77	4550.47	13,176.24
2	65*	122*	8814.00	4407.00	13,221.00*
3	70	117	8907.69	4286.43	13,194.12
4	79.95	110	8961.02*	4091.02	13,052.03
5	85	107	8952.32	4007.97	12,960.29
6	95	102	8899.92	3865.81	12,765.73

standard deviation 30. Note the change of the optimal q and EP_S with the increasing of b, which implies that the supplier obtains the less profit with increasing markdown money. MMP2 case shows the scenario that could achieve the channel coordination. To coordinate the supply chain, in this case, the supply managers should offer $b^*_{SC} = 65$ based on $b^*_{SC} = [(w-c)(r-v)]/(r-c)$.

Note the MMP2 and MMP4, the supplier's profit is larger in the coordinated supply chain than that in the decentralized supply chain in this numerical analysis. On the contrary, the retailer's profit is smaller in the coordinated supply chain according to Proposition 4. In order to clear the relationship of the markdown money b and the expected profit of the supply chain, retailer, and supplier, respectively, we depict the Chart 6.3. Chart 6.3 shows that retailer's profit is always decreasing with b, and supplier's profit is unimodal with b. Therefore, it exits the optimal markdown money which maximizes the channel profit.

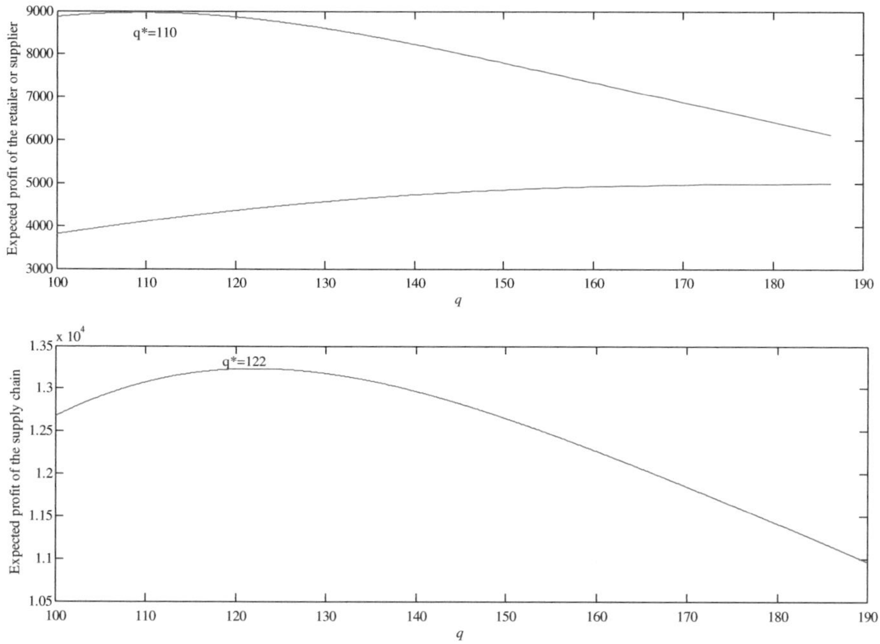

Chart 6.2 Expected profit functions of the retailer, supplier, and supply chain with q

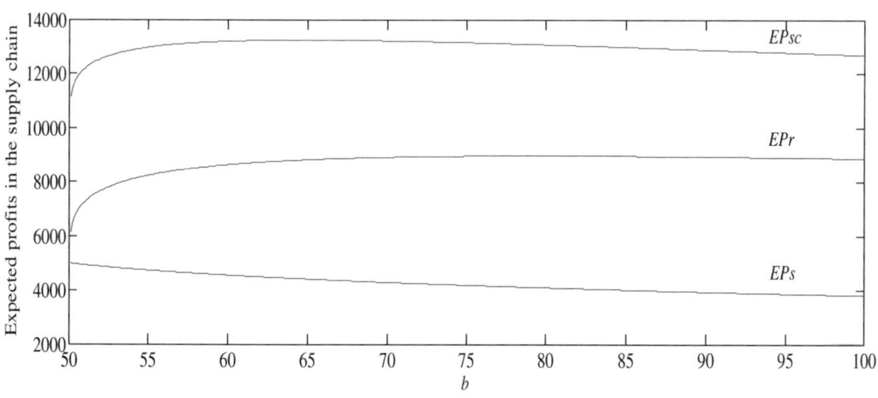

Chart 6.3 Expected profit functions of the retailer, supplier, and supply chain with b

6.5 Conclusion

In this paper, we examine a two-echelon VMI fashion supply chain with one single supplier and one single retailer based on the newsvendor model. We explore how the MMP influences the supply chain under a VMI program in the centralized and decentralized supply chains, respectively. The results demonstrate that similar to a traditional inventory strategy (i.e., RMI), the retailer receives the benefits from the MMP while the supplier suffers from it in both decentralized and centralized scenarios. Therefore, it may not be wise for the supplier to offer markdown money. Furthermore, in the decentralized supply chain, we show that the expected profit of the retailer is not strictly increased by markdown money, but it is always larger than that in the condition without markdown money. We detailed the condition when the optimal markdown money attains the upper bound of it. In addition, if the retailer offers markdown money exceeding the optimal one that should be set, the supplier may correct the markdown money since he could cost less to do this.

In the centralized supply chain, we find that markdown money could allocate the channel profit between the supplier and retailer under the VMI program. This implies that the higher markdown money allows the retailer to receive the higher profit, but the supplier obtains the lower profit. Comparing the profits under the decentralized and the centralized supply chain settings, we know that the total channel profit is higher under the centralized case. Moreover, the supplier's expected profit in a coordinated case is more than that in the decentralized supply chain, while the retailer benefits less when achieving supply chain coordination.

For the future research, it would be interesting to consider the cases that a sales effort is an influencing factor in an overstocked situation within the supply chain (Wong et al. 2009), and how a combination of several contracts improves the channel performance in a VMI partnership (Shen et al. 2014).

Acknowledgments This research is supported in part by: (i) the National Natural Science Foundation of China (71401029); (ii) the Shanghai Pujiang Program (14PJ1400200); (iii) the Fundamental Research Funds for the Central Universities, and Chenguang Program (15CG34).

Appendix: Proofs

Proof of Proposition 1 From (6.1), we have

$$\mathrm{EP_S}(q_\mathrm{S}(b)) = (w-c)q_\mathrm{S}(b) - b\int_0^{q_\mathrm{S}} F(x)\mathrm{d}x$$

$$\begin{aligned}\frac{\mathrm{dEP_S}(b, q_\mathrm{S}(b))}{\mathrm{d}b} &= (w-c)\frac{\mathrm{d}q_\mathrm{S}(b)}{\mathrm{d}b} - \int_0^{q_S} F(x)\mathrm{d}x - bF(q_\mathrm{S}(b))\frac{\mathrm{d}q_\mathrm{S}(b)}{\mathrm{d}b} \\ &= (w-c-bF(q_\mathrm{S}(b)))\frac{\mathrm{d}q_\mathrm{S}(b)}{\mathrm{d}b} - \int_0^{q_\mathrm{S}} F(x)\mathrm{d}x.\end{aligned}$$

Substitute (6.2) in the above equation, and we know $\int_0^{q_S} F(x)\mathrm{d}x > 0$, thus $\mathrm{dEP_S}(q_S(b))/\mathrm{d}b < 0$, that indicates $\mathrm{EP_S}(q_{S^*}(b))$ is decreasing in b, $\forall 0 \le b \le w - v$. When q is optimal quantity $q_{S^*}(b)$, we have

$$\frac{\mathrm{dEP_S}(b, q_{S^*}(b))}{\mathrm{d}q_{S^*}(b)} = (w - c) - b^* F(q_{S^*}(b)) = 0,$$
$$\frac{\mathrm{d}q_{S^*}(b, w)}{\mathrm{d}b} = -\frac{F(q_{S^*}(b, w))}{bf(q_{S^*}(b, w))} < 0.$$

The result shows that $q_{S^*}(b)$ is decreasing in b.

Proof of Proposition 2 For retailer, the expected profit is

$$\mathrm{EP_R}(q_{S^*}(b)) = (r - w)q_{S^*}(b) - (r - b - v)\int_0^{q_{S^*}} F(x)\mathrm{d}x$$

In order to determine the optimal b, we have

$$\frac{\mathrm{dEP_R}(b, q_{S^*}(b))}{\mathrm{d}b} = (r - w)\frac{\mathrm{d}q_{S^*}(b)}{\mathrm{d}b} + \int_0^{q_{S^*}} F(x)\mathrm{d}x - (r - b - v)F(q_{S^*}(b))\frac{\mathrm{d}q_{S^*}(b)}{\mathrm{d}b}$$
$$= [(r - w) - (r - b - v)F(q_{S^*}(b))]\frac{\mathrm{d}q_{S^*}(b)}{\mathrm{d}b} + \int_0^{q_{S^*}} F(x)\mathrm{d}x.$$

We find that if $b = (r - v)(w - c)/(r - c)$, then

$$[(r - w) - (r - b - v)F(q_{S^*}(b))] = 0 \Rightarrow \mathrm{dEP_R}(b, q_{S^*}(b))/\mathrm{d}b > 0.$$

Thus, $b^* > (r - v)(w - c)/(r - c) > (w - c)$ and b^* are in the region $(w - c, w - v]$.

Regarding the first derivation of $\mathrm{EP_R}$ when $b = w - v$, if $\mathrm{dEP_R}(b, q_{S^*}(b))/\mathrm{d}b|_{b=w-v} \ge 0$, then optimal b^* over the upper bound value $w - v$ and therefore $b^* = w - v$; if $\mathrm{dEP_R}(b, q_{S^*}(b))/\mathrm{d}b|_{b=w-v} < 0$, then optimal b^* satisfied $\mathrm{dEP_R}(b, q_{S^*}(b))/\mathrm{d}b = 0$.

Proof of Proposition 3 First, we proof Eq. (6.4). Consider the supply chain's expected profit

$$\mathrm{EP_{SC}}(q_{SC}) = (r - c)q_{SC} - (r - v)\int_0^{q_{SC}} F(x)\mathrm{d}x.$$

Take the derivative of $\mathrm{EP}_{\mathrm{SC}}(q_{\mathrm{SC}})$ with the respect to q_{SC}, we get

$$q_{\mathrm{SC}^*} = F^{-1}\left(\frac{r-c}{r-v}\right).$$

Since the supply chain is coordinated, i.e., $q_{\mathrm{S}^*} = q_{\mathrm{SC}^*}$, we can determine the optimal b

$$b^*_{\mathrm{SC}} = \frac{(w-c)(r-v)}{(r-c)}.$$

Substituting b^*_{SC} and q_{SC^*} into the expected profit of supply chain $\mathrm{EP}_{\mathrm{SC}}(q_{\mathrm{SC}})$, we can find that b^*_{SC} is able to allocate supply chain profit between the retailer and the supplier.

Proof of Proposition 4 Following the proof of Proposition 2, first we consider the case when $b^* = w - v$. In this case, since $b^* - b^*_{\mathrm{SC}} = w - c - [(r-v)(w-c)]/(r-c) = (r-w)(c-v) > 0$ and $\mathrm{EP}_R(b, q_{\mathrm{S}^*}(b))$ is increasing in b in the range of $(w-c, w-v]$. Therefore, the maximum $\mathrm{EP}_{\mathrm{R}}(b, q_{\mathrm{S}^*}(b))$ is always larger than $\mathrm{EP}_{\mathrm{R}}(b^*_{\mathrm{SC}}, q_{\mathrm{SC}^*})$. When b^* is smaller than $w - v$, we have proved that $b^* > b^*_{\mathrm{SC}}$ in the proof of Proposition 2. Thus, the maximum $\mathrm{EP}_{\mathrm{R}}(b, q_{\mathrm{S}^*}(b))$ is greater than $\mathrm{EP}_{\mathrm{R}}(b^*_{\mathrm{SC}}, q_{\mathrm{SC}^*})$.

For the retailer, since that the optimal integrated markdown money is lower than the markdown money decided in the decentralized supply chain. We find that EP_{S} is decreasing in b, when achieves channel coordination.

References

Bernstein F, Chen F, Federgruen A (2006) Coordinating supply chains with simple pricing schemes: the role of vendor-managed inventories. Manage Sci 52(10):1483–1492

Cachon GP (2003) Supply chain coordination with contracts. In: Handbooks in operations research and management science, vol 11, pp 227–339

Çetinkaya S, Lee CY (2000) Stock replenishment and shipment scheduling for vendor-managed inventory systems. Manage Sci 46(2):217–232

Chen H, Chen J, Chen Y (2006) A coordination mechanism for a supply chain with demand information updating. Int J Prod Econ 103:347–361

Chen X, Hao G, Li X, Yiu K (2012) The impact of demand variability and transshipment on vendor's distribution policies under vendor managed inventory strategy. Int J Prod Econ 39 (1):42–48

Choi TM (2013a) Game theoretic analysis of a multi-period fashion supply chain with a risk averse retailer. Int J Inventory Res 2(1–2):63–81

Choi TM (2013b) Carbon footprint tax on fashion supply chain systems. Int J Adv Manuf Technol 68(1–4):835–847

Choi KS, Dai JG, Song JS (2004) On measuring supplier performance under vendor-managed-inventory programs in capacitated supply chains. Manuf Service Oper Manage 6:53–72

Chow PS, Wang YL, Choi TM, Shen B (2015) An experimental study on the effects of minimum profit share on supply chains with markdown contract: risk and profit analysis. Omega—Int J Manage Sci (forthcoming)

Disney SM, Towill DR (2003) Vendor-managed inventory and bullwhip reduction in a two-level supply chain. Int J Oper Prod Manage 23(6):625–651

Dong Y, Xu K (2002) A supply chain model of vendor managed inventory. Transp Res Part E: Logistics Transp Rev 38(2):75–95

Mishra BK, Raghunathan S (2004) Retailer- vs. vendor-managed inventory and brand competition. Manage Sci 50(4):445–457

Nagarajan M, Rajagopalan S (2008) Contracting under vendor managed inventory systems using holding cost subsidies. Prod Oper Manage 17:200–210

Ning Z, Choi TM, Xie C, Xie L, Dai J (2011) Impact of e-marketplace on supply chain under the markdown policy. Supply Chain Manage Int J 16(6):409–418

Shah J, Goh M (2006) Setting operating policies for supply hubs. Int J Prod Econ 100:239–252

Shen B, Choi TM, Wang Y, Lo CK (2013) The coordination of fashion supply chains with a risk-averse supplier under the markdown money policy. IEEE Trans Syst Man Cybern Syst 43(2):266–276

Shen B, Chow PS, Choi TM (2014) Supply chain contracts in fashion department stores: coordination and risk analysis. Math Probl Eng vol 2014, Article ID 954235, 10 p

Tyan J, Wee HM (2003) Vendor managed inventory: a survey of the Taiwanese grocery industry. J Purchasing Supply Manage 9:11–18

Wang CX, Webster S (2009) Markdown money contracts for perishable goods with clearance pricing. Eur J Oper Res 196(3):1113–1122

Wong WK, Qi J, Leung SYS (2009) Coordinating supply chains with sales rebate contracts and vendor-managed inventory. Int J Prod Econ 120(1):151–161

Yao Y, Evers PT, Dresner ME (2007) Supply chain integration in vendor-managed inventory. Decis Support Syst 43(2):663–674

Zhang ML (2010) Revolution! Revolution! The era of zero inventory is coming, assortment upgrades. Garment Dealers, 7 July 2010

Chapter 7
Fashion-Retailing Inventory Management with Random Supply

Qingying Li, Ciwei Dong and Ruixin Zhuang

Abstract We consider a fashion-retailing inventory problem in this chapter. We focus on exploring the scenario in which the fashion retailer, which is modeled as a newsvendor, places an order for a fashionable product to the fashion supplier with random supply. The retail selling price of a fashionable product is a decreasing function of the supplier's random supply. We further consider two scenarios regarding the product cost: (i) the procurement case, in which the fashion retailer is a pure retailer; and (ii) the in-house production case, in which the fashion retailer also manufactures the product by itself. The optimal ordering decisions for both cases are determined and compared. Sensitivity analysis is conducted and insights are revealed.

Keywords Newsvendor · Fashion retailing · Random supply · Supply-dependent price

---

Q. Li
Glorious Sun School of Business and Management, Donghua University,
1882 West Yan'an Road, Shanghai 200051, China
e-mail: liqingying@dhu.edu.cn

C. Dong (✉)
School of Business Administration, Zhongnan University of Economics and Law,
Wuhan 430073, China
e-mail: dongciwei@znufe.edu.cn

R. Zhuang
China Foreign Exchange Trade System, Building 40, 1387 Zhangdong Road,
Shanghai 201203, China
e-mail: zhuangruixin@chinamoney.com.cn

T.-M. Choi (ed.), *Analytical Modeling Research in Fashion Business*,
Springer Series in Fashion Business, DOI 10.1007/978-981-10-1014-9_7

7.1 Introduction

In fashion retailing, a manager's ordering decision can be modeled by a newsvendor problem. It is because that the characteristics of the newsvendor model closely fit the industrial practice of fashion retailing in the following three aspects: high-demand uncertainty, short and single selling season without the replenishment during the season, and relatively simple cost-revenue structure with seldom markdowns. A newsvendor problem is a classical inventory management problem, which has been wildly applied in the studies of the supply chain management in fashion retailing (Choi and Chiu 2012). In a newsvendor setting, the firm aims to decide the amount of products to order for a market with uncertain demand. Interest readers may refer to Khouja (1999), Qin et al. (2011), and Choi (2012) for various newsvendor-based inventory problems. In the traditional newsvendor model, market prices of the products are usually given and are not affected by other factors, such as the supply. Besides, the supply capacity is assumed to be reliable and sufficient in the traditional newsvendor model.

However, in real-world practices of fashion retailing, the supply of the fashion products may be uncertain or even random. For example, cotton is a main raw material of fashion products, and the supply of the cotton-based fashion products may highly depend on the supply of the cotton. Meanwhile, the supply of cotton is closely related to the environment weather condition. The better the weather condition is, the better harvest of the cotton will be, so that the supply will be greater. To effectively match the supply with demand, a fashion-retailing manager needs to make decisions with the consideration of random supply. Meanwhile, the market price of the fashion products will be affected by the supply of the products. When the supply is deficient, the selling price usually goes high; when the supply is sufficient, the selling price may stay low. That is, the market price of the fashion products is a supply-dependent price. Thus, the effects of the random supply and the supply-dependent price cannot be neglected when we consider a newsvendor setting for the fashion retailing. Therefore, we study a newsvendor model in fashion retailing with the consideration of the two above-mentioned characteristics. Through this research, we hope to understand how the fashion retailing manager makes the production order decisions facing a random supply and a supply-dependent price.

7.1.1 Problem Setting

The basic setting of our problem is a newsvendor model. Based on the newsvendor model, we consider the random supply and the supply-dependent retail price. For the market price of the product, we do not make any particular assumption or requirement, except that it decreases in the supply. Based on the common situations in practice, we consider two types of structures for the production cost. The first one is the procurement case, where the production cost depends on the quantity received. The second one is the in-house production case, where the production cost depends on the quantity ordered.

7.1.2 *Overview of Results*

We derive the optimal solutions under each cost structure. Here, we provide an overview of main results:

- The expected profit functions are quasi-concave and concave in the procurement and in-house production cases, respectively. Thus, we have the unique optimal quantity solutions for both cases.
- By comparing the optimal solutions for the two cost structures with that in the traditional newsvendor model, we find that in each case, the optimal order quantity is no greater than that in the traditional newsvendor model with a fixed market price.
- The optimal solution for the procurement case is no smaller than that for the in-house production case.
- By using a numerical example with the settings of the normally distributed random supply and demand, and a linear price function, we investigate the sensitivities of the optimal solutions versus the distribution of the random supply. The results show that the optimal quantities decrease in the deviation of the random supply in both cases, and the optimal quantity of the in-house production case is more sensitive in the deviation of the random supply.

The rest of this chapter is organized as follows. In Sect. 7.2, we review the related literature. In Sect. 7.3, we introduce the model and solve models for the two cases in two subsections. Section 7.4 presents the sensitivity analysis, and Sect. 7.5 concludes the chapter and provides some further research directions. All proofs can be found in Li et al. (2015).

7.2 Literature Review

There are some papers in the literature that study the problems of fashion supply chain management based on the newsvendor models. For example, Choi and Chiu (2012) model a fashion retailer‘s inventory decision making problem as a newsvendor problem and then explore the effect of risk-aversion on the optimal order quantity under the mean-risk framework. Shen et al. (2013) study the use of markdown contract in a fashion supply chain with a risk-averse supplier, where the problem of the fashion retailer is modeled based on the newsvendor problem. Shen et al. (2014) consider a fashion supply chain selling a newsvendor type of fashion product to consumers and investigate the channel coordination issue and examine the performances of different supply contracts. Different from our work, all these papers consider that there is no uncertainty on the supply side.

In our work, we consider that the supply of the fashion product is uncertain. There are three common types of supply uncertainties in the related literature, in terms of the supply quantities. The first one is the supply disruption, where a supplier may not be available to provide the service for random durations. Usually, all ordered items will be delivered if the supplier is available, and nothing will be delivered if the supplier is unavailable. The second one is the random yield, where the quantity received is not the same with the quantity ordered and is usually just a fraction of the quantity ordered. The third one is the random capacity, where the supplier's capacity is a random variable (Tajbakhsh et al. 2007).

Snyder et al. (2014) provide a comprehensive review of the literature on supply disruption. Many prior studies on supply disruption consider the single supplier and focus on using the inventory policy to deal with the supply uncertainty, such as Meyer et al. (1979), Bielecki and Kumar (1988), Parlar and Berkin (1991), Parlar and Perry (1995), Parlar (1997), Moinzadeh and Aggarwal (1997), and Arreola-Risa and DeCroix (1998). Recently, there are some studies on supply disruption considering multiple suppliers and souring mitigation, such as Tomlin (2006), Chopra et al. (2007), Li et al. (2010), Gupta et al. (2014), and Ray and Jenamani (2014).

For the literature of the random yield, a comprehensive review for the random yield can be found in Yano and Lee (1995), and we review some related studies as follows. Kazaz (2004) studies the production planning problem for olive oil with a random yield rate. He considers that both the cost and the price depend on the yield rate. His work is extended by Kazaz and Webster (2011), where the demand is dependent on the yield-rate and risk-averse objective is considered. Xu and Lu (2013) consider a problem in which the manager needs to make a joint decision on the inventory and pricing, based on the newsvendor model. They consider two production cost structures and compare the optimal solutions for the two cases. We adopt these two cost structures used in Xu and Lu (2013), whereas our models are fundamentally different from their model in that we consider a capacity-dependent market price and they consider a fixed market price. Some related papers that study the joint decision of the pricing and inventory policies with random yield include Li and Zheng (2006), Yan and Wang (2013), and Chao et al. (2014). Other recent works about the random yield can be found in Gurnani et al. (2000), Tomlin and Wang (2005), Dada et al. (2007), Giri (2011), Tang and Kouvelis (2011), Chao et al. (2014), and Kazaz and Webster (2015), etc.

Next, we review the third kind of supply uncertainty: random capacity. Arrow et al. (1958) may be considered as the first piece of work on this topic, where they assume that the amount received is independent of the order quantity but depends on a certain probability distribution. Ciarallo et al. (1994) study a multi-period production planning problem with the random capacity. They show that for the single-period model, the random capacity only affects the optimal expected profit but not the optimal order quantity. They then extend the analysis and show that an order-up-to policy, which depends on the capacity distribution, is optimal for the multi-period and infinite-horizon problem. An extended model is studied by Wang and Gerchak (1996), where they consider both random yield and variable capacity. Okyay et al. (2014) study a similar model with dependent demand and supply. Wu et al. (2013) investigate a risk-averse newsvendor model with random capacity, by considering both the risk criteria VaR and CVaR. There are also some papers which study the random capacity without considering the newsvendor model. For example, Bollapragada et al. (2004a) consider an assembly system where the component providers face the random capacity problems, and Erdem et al. (2006) study an EOQ model with multiple suppliers dealing with the random capacity problems. Other studies that consider the random capacity include Khang and Fujiwara (2000), Kouvelis and Milner (2002), Bollapragada et al. (2004b), and Luo et al. (2015). All studies considering the random capacity mentioned above consider a fixed market price.

There are also some more recent newsvendor-based inventory models with supply uncertainty. For example, Ockenfels and Selten (2015) consider the impulse balance and multi-period feedback in the newsvendor model under a game setting. O'Neil et al. (2015) investigate several newsvendor models with demand shocks and unknown demand distributions. Pal et al. (2015) study a newsvendor model with nonlinear holding cost and without specifying the distribution of the demand.

7.3 The Extended Newsvendor Model[1]

As mentioned in the Introduction, a newsvendor model can well fit the fashion retailing model. The newsvendor model is first studied in the literature in 1888 (Chen et al. 2016). In the newsvendor model, there is a single-period selling season. The ordering decision will be made before the selling season, and the demand in the selling season is stochastic. During the selling season, unmet demand is lost, and the left-over quantity can be returned at a salvage value.

The following notations are introduced for the newsvendor model. Let D be the random demand, and let $G(\cdot)$ and $g(\cdot)$ be its *cdf* and *pdf*, respectively. Let c be the unit cost for the product. Unmet demand is lost, and the unit salvage value is s, where $s<c$. The unit selling price of the demand is denoted as p. For notational convenience, we denote $\overline{G}(x) = 1 - G(x)$. Consider the expected profit in the newsvendor model. Let $\pi_0(q)$ be the expected profit given that the order quantity is q, and let q_0^* be the optimal order quantity which maximizes the expected profit. Then, the firm's profit is given by

[1]This chapter's analytical model is based on the article "Managing the newsvendor modeled product system with random capacity and capacity-dependent price" by Li, Q., C. Dong, and R. Zhuang, published 2015 in **Mathematical Problems in Engineering**, http://dx.doi.org/10.1155/2015/296132.

$$\pi_0(q) = (p - s)\left[q - \int_0^q G(\xi)\mathrm{d}\xi\right] - (c - s)q. \tag{7.1}$$

The expected profit $\pi_0(q)$ is shown to be concave, and the optimal order quantity satisfies the following:

$$\overline{G}(q_0^*) = \frac{c - s}{p - s}.$$

The above results about the newsvendor model can be found in many classical textbook, see for example, Zipkin (2000).

In the above introduction of the newsvendor model, we can see that a traditional newsvendor model assumes the supply is reliable/certain, and the selling price is fixed. As mentioned in the introduction section, to better fit the fashion retailing model, we need to introduce the random supply and supply-dependent price into the newsvendor model. As we mentioned above, there are three commonly used supply uncertainty. In this chapter, we restrict our attention to the case with random supply capacity. Without causing any confusion, we simply refer it as "random supply" as follows.

Suppose that the supply is random, denoted as u, and let $f(\cdot)$ and $F(\cdot)$ be its *pdf* and *cdf*, respectively. Similarly, for notational convenience, we let $\overline{F}(x) = 1 - F(x)$. Then, the actually received quantity is the minimum value between the order quantity and the supply, i.e., $\min\{q, u\}$. On the other hand, we also assume that in the selling season, the market price depends on the supply. Specifically, we let the market price be $p(u)$. In our analysis, we only require $p(u)$ satisfies the following two properties:

(i) $p(u)$ decreases in u; that is, the market price decreases with supply.
(ii) $p(u) \geq c$ for any u; that is, the market price is no less than the unit production cost, because, otherwise, the firm will face a nonprofitable market price.

Regarding the fashion retailer's cost for purchasing these $\min\{u, q\}$ units, we will consider two cases. The first case is the procurement case, which is common when the fashion retailer plays a role as a pure retailer. The second case is the in-house production case, which is common when the fashion retailer also plays a role as a manufacturer.

Case 1 is the procurement case. In this case, the fashion retailer needs to pay for only the quantity received, i.e., $\min\{q, u\}$. Consider the expected profit in this case. Let $\pi_{\mathrm{pc}}(u, q)$ be the firm's expected profit given that the supply is u and the order quantity is q. As the received quantity is $\min\{q, u\}$, and the firm only needs to pay for $\min\{q, u\}$ units, $\pi_{\mathrm{pc}}(u, q)$ differs from $\pi_0(\min\{q, u\})$ only in that the market price $p(u)$ is supply-dependent rather than a constant. Replacing "p" in (7.1) by "$p(u)$," we have

$$\pi_{\text{pc}}(u,q) = \begin{cases} (p(u)-s)\left[u - \int\limits_0^u G(\xi)\mathrm{d}\xi\right] - (c-s)u, & \text{if } u<q; \\ (p(u)-s)\left[q - \int\limits_0^q G(\xi)\mathrm{d}\xi\right] - (c-s)q, & \text{if } u \geq q. \end{cases} \tag{7.2}$$

The expected profit can then be obtained by taking expectation on $\pi_{\text{pc}}(u,q)$ over u. Denote $\Pi_{\text{pc}}(q)$ as the fashion retailer's expected profit for the procurement case. We have

$$\Pi_{\text{pc}}(q) = \int\limits_0^{+\infty} \pi_{\text{pc}}(u,q)f(u)\mathrm{d}u.$$

From Li et al. (2015), we know that $\Pi_{\text{pc}}(q)$ is quasi-concave in q. Thus, the optimal order quantity can be uniquely determined by solving

$$\overline{G}(q^*_{\text{pc}}) = \frac{(c-s)\overline{F}(q^*_{pc})}{\int_{q^*_{\text{pc}}}^{+\infty} p(u)f(u)\mathrm{d}u - s\overline{F}(q^*_{pc})}.$$

Case 2 is the in-house production case. In this case, no matter the receiving quantity $\min\{u,q\}$ equals u or q, the fashion retailer, as a manufacturer, needs to pay for the input quantity, i.e., the quantity ordered q.

Similar to the procurement case, we let $\pi_{\text{in}}(u,q)$ be the fashion retailer's profit when the capacity is u and the order quantity is q. When the supply is u, the market price is $p(u)$. In this case, the fashion retailer always needs to pay for the quantity ordered, i.e., the quantity q. Similar to (7.2), we have

$$\pi_{\text{in}}(u,q) = \begin{cases} (p(u)-s)\left[u - \int\limits_0^u G(\xi)\mathrm{d}\xi\right] - (c-s)q, & \text{if } u<q; \\ (p(u)-s)\left[q - \int\limits_0^u G(\xi)\mathrm{d}\xi\right] - (c-s)q, & \text{if } u \geq q. \end{cases}$$

Let $\Pi_{\text{in}}(q)$ be the expected profit for the fashion retailer with in-house production cost. Then,

$$\Pi_{\text{in}}(q) = \int\limits_0^{+\infty} \pi_{\text{in}}(u,q)f(u)\mathrm{d}u.$$

It is known that $\Pi_{\text{in}}(q)$ is the integration of $\pi_{\text{in}}(u, q)$ over u, $\Pi_{\text{in}}(q)$ is concave, and the optimal solution can be uniquely found by solving

$$\overline{G}(q_{\text{in}}^*) = \frac{c - s}{\displaystyle\int_{q_{\text{in}}^*}^{+\infty} p(u)f(u)\mathrm{d}u - s\overline{F}(q_{\text{in}}^*)}.$$

Next, we compare the optimal order quantities in the different cases. Let $q_0^*(x)$ be the optimal order quantity in the traditional newsvendor model given that the market price is x (the traditional newsvendor model considers a fixed market price). We can compare the optimal quantities between the models with two cost structures and also the ones under the traditional newsvendor model. The result is shown in Theorem 7.1.

Theorem 7.1 $q_0^*(p(q_{pc}^*)) \geq q_{pc}^* > q_{in}^*$.

Ciarallo et al. (1994) shows that in the traditional newsvendor model with a fixed market price, the random supply does not affect the optimal order quantity. However, the first inequality in Theorem 7.1 shows that this is not valid when the market price is also supply-dependent. In the procurement model, where the market price varies with the random supply, the optimal order quantity q_{pc}^* is no greater than the optimal order quantity in the traditional newsvendor model with a fixed market price $p(q_{\text{pc}}^*)$.

The second inequality in Theorem 7.1 shows that the optimal order quantity in the in-house production case is no greater than the optimal order quantity in the procurement case. This is consistent with the intuitive belief. In the in-house production case, the fashion retailer always needs to pay for the quantity ordered, even if it is smaller than the quantity received. However, in the procurement case, the fashion retailer only needs to pay for the quantity received. Thus, the fashion retailer facing in-house production cost has no incentive to order as high as the one facing procurement case.

7.4 Sensitivity Analysis

To investigate the sensitivity of the optimal solution versus the random supply, we now provide the sensitivity analysis. Let the market demand and the random supply both follow a normal distribution with mean $\mu = 5$ and standard deviation $\sigma = 1$, let the unit ordering cost $c = 2$, and let the salvage value $s = 0.2$. In the numerical example, we will set an error tolerance level = 1×10^{-4}. With this tolerance level, we can restrict our attention to the random supply with a domain of $[0, 10]$. We use the price function $p(u) = 4 - 0.2u$.

Table 7.1 The sensitivity analysis

	$q_0^*(\cdot)$	$q_{\text{pc}}^*(\cdot)$	$q_{\text{in}}^*(\cdot)$	Relationship
$\sigma_u \uparrow$	Constant	Insensitive, decrease	Sensitive, decrease	$q_0^*(\sigma_u) \geq q_{\text{pc}}^*(\sigma_u) > q_{\text{in}}^*(\sigma_u)$
$\sigma_d \uparrow$	Insensitive, decrease	Insensitive, decrease	Sensitive, decrease	$q_0^*(\sigma_d) \geq q_{\text{pc}}^*(\sigma_d) > q_{\text{in}}^*(\sigma_d)$

The sensitivity analysis of the optimal solutions is conducted versus the random demand and the random supply, respectively. When we conduct the sensitivity analysis versus the random supply u, we consider a fixed distribution of the random demand and let the deviation of the random supply, denoted as σ_u, varies from 0 to 1.2. Similarly, when we conduct the sensitivity analysis versus the random demand d, we consider fixed distribution of the random supply and let the deviation of the random demand, denoted as σ_d, varies from 0 to 1.2.

For each σ_u and σ_d, we calculate the optimal order quantities for the two cases as well as the optimal order quantity for the traditional newsvendor model with market price equaling $\mathbb{E}p(u) = 3$. The results are summarized in Table 7.1. For the details, please refer to Li et al. (2015).

The row with σ_u in Table 7.1 shows the sensitivity result versus random supply. The first result is that the optimal order quantities for both of the two cases decrease as σ_u increases. This means that when the random supply has greater deviation, the fashion retailer has the incentive to order less. Another result is that the decrease of the optimal quantity for the in-house production case is sharper than the decrease of the optimal quantity for the procurement case. This is because in the in-house production case, the fashion retailer has to pay for the quantity ordered. Compared with the procurement case, the in-house production case's expected profit is affected more by the random supply. Thus, for a fashion retailer where the in-house production model is adopted, the manager should be more careful when the supply has greater deviation.

The row with σ_d in Table 7.1 shows the sensitivity result versus the random demand. In this case, the optimal order quantities in the three cases all decrease when the standard deviation increases, the optimal order quantity in the in-house production case is always the smallest, and the optimal order quantity in the traditional newsvendor model is always the greatest. We also observe that optimal order quantity in the in-house production case decreases sharper than that in the procurement case or the traditional newsvendor case. Therefore, for a firm adopting the in-house production purchasing cost structure, the manager shall be more careful about the volatility of the demand and the supply.

7.5 Conclusion and Further Research

We have studied an extended newsvendor model with a random supply and the supply-dependent market price for fashion business. Two production cost structures are considered: the procurement cost structure where the fashion retailer pays for the received quantity, and the in-house production cost structure where the fashion retailer pays for the input quantity. The market price depends on the random supply, and we consider a general form only requiring that the price decreases in the supply. We show that the optimal quantity is unique and specify the optimal conditions under both cost structures. We also find that the optimal solution for the procurement case is no smaller than that for the in-house production case. Sensitivity studies are also conducted. We observe that the optimal quantity decreases in the deviation of the random supply/demand in both of the two production cases, and that the optimal quantity of the in-house production case is more sensitive in the deviation of the random supply/demand. The fashion retailer managers should be more careful about the volatility of the random supply/demand if the system is adopting an in-house production purchasing cost structure.

Further research may include several directions. The first one is the stochastic price-dependent demand process. Here, the demand process and the price are independent. In many fashion business situations, it is common that the price will highly affect the demand. Therefore, the price-dependent demand process is worthy of further investigation. Second, what we considered is a risk-neutral decision maker who maximizes the expected profit. Many practical business managers may hate profit loss than profit gain. Such managers are risk averse, and they may take VaR, CVaR, or profit variance as the risk-averse objective. Third, we assume that the random supply and the demand are two independent processes. It might be more practical if we allow the two random variables to be correlated. However, this would greatly increase the difficulty of solving the problem. We take it as an interesting further research topic.

Acknowledgments The authors would like to thank Prof. Tsan-Ming Choi for the helpful suggestions and comments. This research was supported in part by National Natural Science Foundation of China (71501037). It was also supported in part by the Fundamental Research Funds for the Central Universities.

References

Arreola-Risa A, DeCroix GA (1998) Inventory management under random supply disruptions and partial backorders. Naval Res Logistics 45:687–703

Arrow KJ, Karlin S, Scarf H (1958) Studies in the mathematic theory of inventory and production. Stanford University Press, Stanford, CA

Bielecki T, Kumar PR (1988) Optimality of zero-inventory policies for unreliable manufacturing systems. Oper Res 36(4):532–541

Bollapragada R, Rao US, Zhang J (2004a) Managing inventory and supply performance in assembly systems with random supply capacity and demand. Manage Sci 50(12):1729–1743

Bollapragada R, Rao US, Zhang J (2004b) Managing two-stage serial inventory systems under demand and supply uncertainty and customer service level requirements. IIE Trans 36 (1):73–85

Chao X, Gong X, Zheng S (2014) Optimal pricing and inventory policies with reliable and random-yield suppliers: characterization and comparison. Ann Oper Res. doi:10.1007/s10479-014-1547-0

Chen R, Cheng TCE, Choi TM, Wang Y (2016) Novel advances in applications of the newsvendor model. Decis Sci

Choi TM (2012) Handbook of newsvendor problems. Springer, New York

Choi TM, Chiu CH (2012) Mean-downside-risk and mean-variance newsvendor models: Implications for sustainable fashion retailing. Int J Prod Econ 135:552–560

Chopra S, Reinhardt G, Mohan U (2007) The importance of decoupling recurrent and disruption risks in a supply chain. Naval Res Logistics 54(5):544–555

Ciarallo FQ, Akella R, Morton TE (1994) A periodic review, production planning model with uncertain capacity and uncertain demand—optimality of extended myopic polices. Manage Sci 40:320–332

Dada M, Petruzzi NC, Schwarz LB (2007) A newsvendor's procurement problem when suppliers are unreliable. Manuf Serv Oper Manage 9:9–32

Erdem AS, Fadiloğlu MM, Özekici S (2006) An EOQ model with multiple suppliers and random capacity. Naval Res Logistics 53(1):101–114

Giri BC (2011) Managing inventory with two suppliers under yield uncertainty and risk aversion. Int J Prod Econ 133:80–85

Gupta G, He B, Sethi SP (2014) Contingent sourcing under supply disruption and competition. Int J Prod Res. http://dx.doi.org/10.1080/00207543.2014.965351

Gurnani H, Akella R, Lehoczky J (2000) Supply management in assembly systems with random yield and random demand. IIE Trans 32(8):701–714

Kazaz B (2004) Production planning under yield and demand uncertainty with yield-dependent cost and price. Manuf Serv Oper Manage 6:209–224

Kazaz B, Webster S (2011) The impact of yield-dependent trading costs on pricing and production planning under supply uncertainty. Manuf Serv Oper Manage 13:404–417

Kazaz B, Webster S (2015) Technical note—price-setting newsvendor problems with uncertain supply and risk aversion. Oper Res 63(4):807–811

Khang DB, Fujiwara O (2000) Optimality of myopic ordering policies for inventory model with stochastic supply. Oper Res 48(1):181–184

Khouja M (1999) The single-period (news-vendor) problem: Literature review and suggestions for future research. Omega 27(5):537–553

Kouvelis P, Milner JM (2002) Supply chain capacity and outsourcing decisions: the dynamic interplay of demand and supply uncertainty. IIE Trans 34(8):717–728

Li Q, Zheng S (2006) Joint inventory replenishment and pricing control for systems with uncertain yield and demand. Oper Res 54:696–705

Li J, Wang S, Cheng TCE (2010) Competition and cooperation in a single-retailer two-supplier supply chain with supply disruption. Int J Prod Econ 124:137–150

Li Q, Dong C, Zhang R (2015) Managing the newsvendor modeled product system with random capacity and capacity-dependent price. Math Prob Eng 2015:10 (Article ID 296132). http://dx.doi.org/10.1155/2015/296132

Luo Z, Wang J, Chen W (2015) A risk-averse newsvendor model with limited capacity and outsourcing under the CVaR criterion. J Syst Sci Syst. http://dx.doi.org/10.1007/s11518-015-5263-3

Meyer RR, Rothkopf MH, Smith SA (1979) Analysis of a production/inventory system subject to random disruptions. Manage Sci 43(11):1577–1588

Moinzadeh K, Aggarwal P (1997) Reliability and inventory in a production-storage system. Manage Sci 25(8):799–807

Ockenfels A, Selten R (2015) Impulse balance and multiple-period feedback in the newsvendor game. Prod Oper Manage

Okyay HK, Karaesmen F, Ozekici S (2014) Newsvendor models with dependent random supply and demand. Optim Lett 8:983–999

O'Neil S, Zhao X, Sun D, Wei JC (2015) Newsvendor problems with demand shocks and unknown demand distributions. Decis Sci

Pal B, Sana SS, Chaudhuri K (2015) A distribution-free newsvendor problem with nonlinear holding cost. Int J Syst Sci. http://dx.doi.org/10.1080/00207721.2013.815828

Parlar M (1997) Continuous-review inventory problem with random supply interruptions. Eur J Oper Res 99:366–385

Parlar M, Berkin D (1991) Future supply uncertainty in EOQ models. Naval Res Logistics 38:107–121

Parlar M, Perry D (1995) Analysis of a (*Q*, *r*, *T*) inventory policy with deterministic and random yields when future supply is uncertain. Eur J Oper Res 84:431–443

Qin Y, Wang R, Vakharia AJ, Chen Y, Seref MMH (2011) The newsvendor problem: review and directions for future research. Eur J Oper Res 213(2):361–374

Ray P, Jenamani M (2014) Sourcing decision under disruption risk with supply and demand uncertainty: a newsvendor approach. Ann Oper Res. doi:10.1007/s10479-014-1649-8

Shen B, Choi TM, Wang Y, Lo CK (2013) The coordination of fashion supply chains with a risk-averse supplier under the markdown money policy. IEEE Trans Syst Man Cybern: Syst 43 (2):266–276

Shen B, Chow PS, Choi TM (2014) Supply chain contracts in fashion department stores: coordination and risk analysis. Math Probl Eng. http://dx.doi.org/10.1155/2014/954235

Snyder LV, Atan Z, Peng P, Rong Y, Schmitt AJ, Sinsoysal B (2014) OR/MS models for supply chain disruptions: a review. Working paper, Lehigh University, Bethlehem, PA, USA

Tajbakhsh MM, Zolfaghari S, Lee GG (2007) Supply uncertainty and diversification: a review. In: Jung H, Jeong B, Chen FF (eds) Trends in supply chain design and management. Technologies and methodologies. Springer, London

Tang SY, Kouvelis P (2011) Supplier diversification strategies in the presence of yield uncertainty and buyer competition. Manuf Serv Oper Manage 13(4):439–451

Tomlin B (2006) On the value of mitigation and contingency strategies for managing supply chain disruption risks. Manage Sci 52(5):639–657

Tomlin B, Wang Y (2005) On the value of mix flexibility and dual sourcing in unreliable newsvendor networks. Manuf Serv Oper Manage 7(1):37–57

Wang YA, Gerchak Y (1996) Periodic review production models with variable capacity, random yield, and uncertain demand. Manage Sci 42:130–137

Wu M, Zhu SX, Teunter RH (2013) The risk-averse newsvendor problem with random capacity. Eur J Oper Res 231:328–336

Xu M, Lu Y (2013) The effect of supply uncertainty in price-setting newsvendor models. Eur J Oper Res 227:423–433

Yan X, Wang Y (2013) An EOQ model for perishable items with supply uncertainty. Math Probl Eng. http://dx.doi.org/10.1155/2013/739425

Yano CA, Lee HL (1995) Lot sizing with random yield: a review. Oper Res 43(2):311–334

Zipkin PH (2000) Foundations of inventory management. McGraw-Hill, Boston, MA

Part III
Engineering Models, Applications and Cases

Chapter 8
Tales of a Fashion So(u)rcerer: Optimal Sourcing, Quotation, and In-House Production Decisions

Tarkan Tan and Osman Alp

Abstract Most companies in fashion industry, as well as many other industries, must procure items necessary for their businesses from outside sources, where there are typically a number of competing suppliers with varying cost structures, price schemes, and capacities. This situation poses some interesting research questions from the outlook of different parties in the supply chain. We consider this problem from the perspective of (i) the party that needs to outsource, (ii) the party that is willing to serve as the source, and (iii) the party that has in-house capability to spare. We allow for stochastic demand, capacitated facilities (in-house and suppliers'), and general structures for all relevant cost components. Some simpler versions of this problem are shown to be NP-hard in the literature. We make use of a dynamic programming model with pseudo-polynomial complexity to address all three perspectives by solving the corresponding problems to optimality. Our modeling approach also lets us analyze different aspects of the problem environment such as pricing schemes and channel coordination issues.

Keywords Sourcing · Supplier selection · Inventory · Production · General costs · Capacity · Supply chain · Channel coordination · Fashion industry

---

Parts of Sects. 8.1, 8.2, and 8.5 of this chapter are taken from Tan and Alp (2016).

T. Tan (✉)
School of Industrial Engineering, Eindhoven University of Technology,
P.O. Box 513, 5600 MB Eindhoven, The Netherlands
e-mail: t.tan@tue.nl

O. Alp
Haskayne School of Business, University of Calgary,
Calgary, AB T2N 1N4, Canada
e-mail: osman.alp@ucalgary.ca

T.-M. Choi (ed.), *Analytical Modeling Research in Fashion Business*,
Springer Series in Fashion Business, DOI 10.1007/978-981-10-1014-9_8

8.1 Introduction and Related Literature

Fashion industry is very challenging in terms of production and inventory planning due to its two main characteristics: (1) the product life cycle is fairly short—especially for fads—because of changing tastes and trends. This typically means that most—if not all—of the production needs to take place before the demand materializes and there is hardly any possibility to adjust the order quantities, considering relatively long production lead times associated with designing the products, sourcing the materials, converting them into final products, and distributing the products to the market. (2) The demand is highly volatile and difficult to predict in advance. This necessitates a careful trade-off between all cost parameters regarding sourcing, manufacturing, underage, and overage in optimizing the production quantities, underage typically being quite costly in the long run. Consequently, an analytical treatment of production and inventory planning for fashion industry incorporating these characteristics is necessary. While the problem can be considered as a basic one in operations management literature characterized by the newsvendor model, the sourcing aspect which is extremely common in today's complex supply chains is mostly overlooked due to the difficulties discussed later in this chapter. In what follows we address this problem from different angles. Since our work does not only apply to fashion industry but also to other industries that are subject to short life cycles and stochastic demand, we follow a generic terminology in the rest of the chapter (i.e., product, component, and manufacturer) instead of one that refers specifically to fashion industry.

Consider a manufacturer or retailer who procures (or, 'sources') a certain product or service, to use directly or indirectly in meeting the stochastic demand that she faces. Considering the manufacturing environment as an example, the product that is to be procured (or, the 'item') can be supplied by a finite number of capacitated external suppliers, and the manufacturer must decide which of the sources to utilize and to what extent. One could prefix the procurement quantity based on inventory- and production-related costs and then find the least costly solution from the available pool of suppliers with corresponding price structures and capacities. However, the optimal sourcing (procurement) decision under stochastic demand requires an integrated approach, using all of the cost parameters and capacity and price information of alternative suppliers simultaneously.

Supplier price and capacity information could be collected by making use of e-business infrastructure or organized industrial associations, or by contacting qualified suppliers, using a request for quotations (RFQ). These sources may have different capacities and price structures, but we consider them to be identical in terms of their function, i.e., the item's characteristics do not depend on the supplier. We do not restrict our analysis to a particular cost function for procurement, and we allow, for example, for a separate fixed cost for initiating the use of each source, for logistics costs that might depend on the geographical location of the suppliers, and for nonlinear unit variable costs. Progressive or all-units quantity discounts are special cases (see Andrade-Pineda et al. 2015, for a particular treatment of a nonlinear quantity discount cost scheme in supplier selection problem). Moreover, the 'cost-of-doing business' with each supplier might incur nonlinear cost factors (Kostamis et al. 2009). The suppliers' capacity utilization might result in re-evaluating the remaining available capacities, inducing quantity-dependent price quotations.

Purchasing is a common operation for all types of businesses. Kaplan and Sawhney (2000) analyze business-to-business e-commerce marketplaces and classify the purchasing market as manufacturing inputs and operating inputs, in terms of what businesses buy and as systematic sourcing and spot sourcing, in terms of how they buy. Our approach applies to any type of manufacturing or operating inputs that face stochastic demand and that are purchased from the spot market: the 'exchanges' and 'yield managers,' respectively (Kaplan and Sawhney 2000). There are numerous Web-based platforms on the market that can materialize the sourcing methodologies prescribed in this study. There are general-purpose B2B e-commerce platforms such as Ariba (2014), Fiatech (2014), and 1 Point Commerce (2014) and specific platforms operated by companies for their operations such as the ones by Ford (2014), Foster Wheeler (2014), and Hilton (2014).

We consider three problems (or sorcerer's tales) in such an environment:

1. Manufacturer's Sourcing Problem: 'How to do the trick?' That is, which sources should be utilized to what extent?
2. Supplier's Problem: 'How to cast a counter-spell?' That is, if a new supplier that intends to bid on the RFQ has information on his competitors' price and capacities, what is the best price he should quote and what is the capacity he should dedicate?
3. Manufacturer's In-House Production Capacity Problem: 'Cast your own spell.' That is, if the manufacturer can allocate some of her resources for in-house manufacturing of the item, how much capacity should she dedicate for this?

We note that our problem environment is extremely general and is not necessarily confined to procurement of goods in fashion industry or even a supply chain context. To name some other environments, consider transportation logistics, manufacturing options, carbon offsetting, and the make-or-buy problem. As for transportation logistics, suppose that the materials ordered by a manufacturer or a retailer are shipped by vehicles with certain capacities. For each vehicle utilized, there may exist a fixed cost as well as a unit variable cost and possibly quantity discounts. The total order may be satisfied with a number of vehicles with varying characteristics. As for the manufacturing options, consider a heating process using industrial ovens. Each oven may have a different capacity and a particular cost of operation, including fixed costs. Similarly, consider a production environment with flexible and dedicated machines, in which each machine incurs different setup and production costs. As for carbon offsetting, consider a socially responsible company that wants to offset its carbon emissions by investing in carbon abatement projects. The company must choose the 'best' (cost minimizing or utility maximizing) way of offsetting, from a number of certified offsetting options with different cost parameters (or utilities) and carbon abatement capacities.

Procurement decisions should consider the cost of materials procured, delivery punctuality, the quality of items procured, creation of effective strategic partnerships, possibly the carbon footprint, and the like. Therefore, one of the key processes of effective supply chain management is the supplier selection process, which consists of determining a supplier base (a set of potential suppliers to operate with), the supplier(s) to procure from, and the procurement quantities. We refer the reader to Elmaghraby (2000) for an overview of research on single- and multiple-sourcing strategies. Aissaouia et al. (2007) present a comprehensive review of literature related to several aspects of the procurement function, including the supplier selection process and in-house versus outsourcing decisions. Firms sometimes employ multiple criteria in selecting their suppliers (Ustun and Demirtas 2008; Hosseininasab and Ahmadi 2015). A survey of multi-criteria approaches for supplier evaluation and selection processes is presented by Ho et al. (2010). More recently, Kumar et al. (2014) introduce a supplier selection approach taking carbon footprint of the suppliers into account. Jia et al. (2015) consider a broader perspective by taking all three aspect of the triple bottom line into account, that is, people, planet, profit in fashion industry. In our work, we do not include the multi-criteria supplier evaluation phase. We assume that the supplier base has already been determined and that the immediate supplier selection decisions are based on the cost criterion.

In our analysis, we consider a single-period, single-item make-to-stock setting. The procurement problem has received much attention, mostly under the deterministic demand assumption (which results in a preset total procurement quantity). When the demand is deterministic, the problem becomes either (i) to determine the set of suppliers to purchase a given quantity, or (ii) to determine the suppliers and the purchasing frequency for a given demand rate. Chauhan and Proth (2003) consider a version of the problem, in which there is a lower and an upper bound for the capacity of each supplier, and the supply costs are concave. They propose heuristic algorithms. Chauhan et al. (2005) show that the problem considered by Chauhan and Proth (2003) is NP-hard. Burke et al. (2008a) consider this problem under different quantity discount schemes and capacitated suppliers. They propose heuristic algorithms to solve the problem. Burke et al. (2008b) discuss that this particular problem is a version of the 'continuous knapsack problem,' in which the objective is to minimize the sum of separable concave functions, and show that this problem is NP-hard. Romeijn et al. (2007) analyze the continuous knapsack problem with nonseparable concave functions and propose a polynomial time algorithm. We note that the supplier selection problem with stochastic demand results in a nonseparable cost function; it is actually not a knapsack problem, because the size of the knapsack (the amount allocated to the suppliers) is itself a decision variable. We provide an exact pseudo-polynomial algorithm to solve the stochastic version of this problem, while not imposing restrictions on the supply cost. We refer the interested reader to Burke et al. (2008b) for a further review of the related literature and to Qi (2007), Kawtummachai and Hop (2005), and Mansini et al. (2012) for different aspects of the problem under deterministic demand. In this study, we contribute to the literature by considering stochastic demand and by including general cost structures.

The stochastic demand version of the procurement problem under capacitated suppliers has also received attention to a certain extent in the literature. Alp and Tan (2008) and Tan and Alp (2009) analyze the problem with two supply options, in a multi-period setting under fixed costs of procurement. Alp et al. (2014) consider an infinite horizon version of this problem with identical suppliers and a linear cost function with a fixed component, which is a special case of ours. Awasthi et al. (2009) consider multiple suppliers that have minimum order quantity requirements and/or a maximum supply capacity, but no fixed cost is associated with procurement. They show that this problem is NP-hard, even when the suppliers quote the same unit price to the manufacturer and propose a heuristic algorithm for the general version. Hazra and Mahadevan (2009) analyze an environment in which the buyer reserves capacity from a set of suppliers through a contracting mechanism. The capacity is reserved before the random demand is observed and allocated uniformly to the selected suppliers. If the capacity is short upon demand realization, the shortage is fulfilled from a spot market at a higher unit price. Our work differs from these articles, because we consider multiple suppliers and general cost functions, and we do not impose a particular structure on the allocation of purchased quantity to the suppliers.

Zhang and Zhang (2011) consider a similar environment to ours. A single item that faces stochastic demand is procured from potential suppliers that have minimum and maximum order sizes, and a fixed procurement cost is considered. They propose a nonlinear mixed-integer programming formulation and a branch-and-bound algorithm. Our problem is more general than this, as we do not impose restrictions on the supply cost structures, a situation that cannot be handled by the methodology proposed by the aforementioned authors. Finally, we note that Zhang and Ma (2009) also consider a similar problem for multiple items. They assume that suppliers are capacitated and offer quantity discounts. A mixed-integer nonlinear programming formulation that determines the optimal production quantities of each product, purchasing quantities of the raw materials, and the corresponding suppliers to make the purchases is proposed. Ayhan and Kilic (2015) propose a two-stage approach to select suppliers under quantity discounts where the first stage is used to find the relative weights of the selection criteria and the second stage selects best suppliers via a MILP model. Another work that is related to our problem environment, particularly considering the problem from the suppliers' point of view, is by Li and Debo (2009). The authors consider an existing and an entrant supplier that compete for the business of the manufacturer. Using a unit variable cost structure and considering a two-period setting where the demand in the second period is stochastic, the authors derive several managerial insights regarding the capacity investment and price quotation decisions of both suppliers.

Our study also elaborates on the value of coordinating the business channel between a supplier and the manufacturer. Several mechanisms such as contracting, quantity discounts, return options have been proposed in the literature in order to coordinate the channel and create a win–win situation. Li and Wang (2007) present a comprehensive review of the channel coordination literature. Toptal and Cetinkaya (2008) quantify the value of channel coordination between a supplier and a buyer under a certain cost structure. Kheljani et al. (2009) consider a buyer's sourcing decisions by focusing on optimizing the channel's profit. Both of these studies consider deterministic demand. Xia et al. (2008) consider the channel coordination problem for a multiple supplier and multiple buyer setting. The order quantity and frequency of the buyers are exogenous parameters. The authors present models that can be used to coordinate the channel by matching the suppliers' cost functions and the buyers' purchasing behaviors.

In the third subproblem, we show how our main methodology can be used to find the optimal in-house production versus outsourcing decision (considering the cost aspect of the problem in isolation), as in-house production can be considered one of the available sourcing options. In such situations, it is likely that the total cost of allocating some or all in-house capacity for producing the item would have a nonlinear nature, stemming from cost components such as fixed costs, incremental capacity usage costs, and concave or convex capacity allocation (opportunity) costs. The complexity of in-house capacity costs is also illustrated by a Darden School of Business case on Emerson Electric Company (Davis and Page 1991). The flexibility of our proposed methodology in its ability to handle all kinds of cost functions is one of our major contributions to literature.

The major contributions of our paper can be summarized as follows:

- We build a novel dynamic programming model that we use for finding the optimal solution to the NP-hard sourcing problem under a fairly general setting consisting of stochastic demand, general cost structures and capacitated suppliers in one shot. The computational complexity of the solution that we propose is pseudo-polynomial.
- We evaluate the performance of decoupling sourcing and production decisions.
- We develop a methodology to find the optimal pricing decision of a supplier who competes with other suppliers.
- We develop a methodology to find the optimal capacity allocation decision of the manufacturer for in-house manufacturing under the existence of alternative production sources.
- Finally, we make observations and build managerial insights, some of which are contrary to the collective intuition that traditional inventory/production models generate.

The rest of the paper is organized as follows: We present the manufacturer's sourcing problem in Sect. 8.2. The supplier's problem is analyzed in Sect. 8.3 and the manufacturer's in-house production capacity problem is analyzed in Sect. 8.4. We conclude the paper in Sect. 8.5.

8.2 Manufacturer's Sourcing Problem: How to Do the Trick?

In this section, we analyze the procurement problem in a single-period setting, under a given set of alternative capacitated suppliers, with corresponding general procurement cost functions. The procured quantity also dictates the stock quantity, subject to stochastic demand. There are two decisions in such an environment: Which sources should be utilized and in what quantities? The relevant parameters in determining those quantities are not only procurement costs and supplier capacities, but also the inventory-related cost parameters in the system. Nevertheless, one could either prefix the total order quantity and then decide on the allocation of this to the supplier base in a sequential manner, or make those decisions in an integrated fashion. The former could be a result of factors such as (i) the perception that procurement-related (external) parameters and production/inventory-related (internal) parameters need to be treated separately; (ii) the time lag between those decisions, e.g., the production department determines required quantities and relays this information to the purchasing department, who makes the purchase with the least cost; (iii) lack of sufficient coordination between separate departments within the organization, e.g., making their uncoordinated decisions based on sales targets and forecasts of the company or their separate performance incentives; (iv) the conventional market and/or company practice of tendering for bids based on a prefixed quantity; (v) lack of sufficient information on the supplier base; and (vi) managerial overlook on the potential savings of integration. In the absence of such factors, solving the problem by considering all problem parameters in an integrated way constitutes the basic research question that we address.

In what follows, we first highlight a major drawback of the sequential approach. Then, we present a dynamic programming model to formulate the problem under consideration and show how the optimal solution can be found in an integrated manner. Finally, we present the results of the numerical study we conducted to investigate (i) the effect of problem parameters on the optimal solution, and (ii) the performance of the sequential approach.

The relevant costs in our environment are the costs of procuring from suppliers and underage and overage costs, all of which are exogenously determined and nonnegative. We do not impose any conditions on the costs of procuring from suppliers, and, hence, these costs might assume any form, including fixed costs for procurement, stepwise costs for shipments, costs that imply minimum order quantities, and different forms of quantity discounts. Our approach allows for the underage and overage costs of the remaining inventory level after demand materialization to also assume any form, via the corresponding loss function. We consider capacitated suppliers with fixed and known capacities. We assume full availability of the ordered quantities, and we also assume that the differences between procurement lead times from alternative suppliers can be neglected. In case the latter assumption is significantly violated, different lead times can be approximately incorporated into the model, by considering appropriate costs associated

Table 8.1 Summary of notation

N	Number of alternative suppliers
Q	Total procurement quantity
U_n	Capacity of supplier $n, n = 1, 2, \ldots, N$
q_n	Quantity procured from supplier n
$C_n(q_n)$	Cost of procuring q_n units from supplier $n, n = 1, 2, \ldots, N$
h	Overage cost per unit unsold
b	Underage cost per unit of unmet demand
W	Random variable denoting the demand
$G(w)$	Distribution function of W

with purchasing from each supplier, reflecting the cost effect of corresponding procurement lead times. Similarly, other non-biddable price factors, such as delivery punctuality, the quality of items procured, and strategic partnership concerns, are also valuated by the manufacturer and reflected in the procurement costs. Naturally, the more differences in non-biddable price factors, the less accurate the cost-based methods (like ours). For a discussion on the valuation of non-biddable price factors, see Kostamis et al. (2009). We summarize our major notation in Table 8.1.

If q_n units are procured from supplier $n, n = 1, 2, \ldots, N$, with a corresponding cost of $C_n(q_n)$, then the total cost of procuring $Q = \sum_n q_n$ units is $PC(Q) = \sum_n C_n(q_n)$, and the resulting average unit procurement cost is $c = PC(Q)/Q$. The problem is to minimize expected total costs $ETC(Q) = PC(Q) + \mathcal{L}(Q)$, where $\mathcal{L}(Q)$ denotes the total expected overage and underage costs, the standard loss function $\mathcal{L}(Q) = h\int_0^Q (Q - w)dG(w) + b\int_0^\infty (w - Q)dG(w)$ being a special case. In the sequential approach, the total order quantity Q^o is decided without knowing the total cost of procurement. This is because it is unknown, a priori, what the exact allocation of the total order quantity to the supplier base is, or whether the supplier base has the total capacity to meet this order. Once the total order quantity is determined, the allocation is optimized by solving the following problem (P), based on the sales prices and capacities quoted by various suppliers:

$$\begin{aligned} \text{Min.} \quad & \sum_n C_n(q_n) \\ \text{st} \quad & \sum_n q_n = \min\left\{Q^o, \sum_n U_n\right\} \\ & q_n \le U_n \quad \text{for all } n. \end{aligned}$$

Note that Q^o is not necessarily equal to the optimal procurement quantity, $\widehat{Q} = \sum_n q_n$. As to the determination of the total order quantity Q^o, if only the inventory-related costs are considered, then the optimal order quantity is $\widehat{Q}^o = \arg, \min_Q\{\mathcal{L}(Q)\}$. But this approach results in overestimation of the required quantity, as it neglects procurement costs. If one prefers to incorporate a linear unit procurement cost of c, the resulting optimal order quantity would be $\widehat{Q}^o(c) = \arg,\min_Q\{cQ + \mathcal{L}(Q)\}$ (in case of standard loss function $\mathcal{L}(Q)$, the solution would then be $\widehat{Q}^o(c) = G^{-1}\left(\frac{b-c}{b+h}\right)\Big)$. However, in general, there is no way of knowing what the actual procurement cost will be, until the required quantity is known. One could prepare a list of all possible quantities, but each entry in the list requires solving problem P, which is a knapsack problem with a general objective function. A special case is the fixed-charge continuous knapsack problem (see Haberl 1999), which is NP-hard with some known pseudo-polynomial algorithms.

A simple approach is to incorporate an estimate of the purchasing cost, $\tilde{c} = \frac{\sum_n C_n(U_n)}{\sum_n U_n}$, and decide on $\widehat{Q}^o(\tilde{c})$ accordingly, after which $\widehat{Q} = \min\left\{\widehat{Q}^o(\tilde{c}) \sum_n U_n\right\}$ units are procured by solving problem P. Nevertheless, this approach can be improved: Once the optimal cost of procuring $\widehat{Q}$ and the corresponding average unit procurement cost $c = PC(\widehat{Q})/\widehat{Q}$ are known, $\widehat{Q}^o$ can be updated by making use of this information, and so forth. Exploiting this idea, one can come up with the following algorithm (where Step 0 makes use of the computations stated above as the simple approach):

Step 0. Set $i = 1$, $\widehat{Q}_i = \min\left\{\widehat{Q}^o(\tilde{c}), \sum_n U_n\right\}, c_{i+1} = PC(\widehat{Q}_i)/\widehat{Q}_i$.

Step 1. Set $i = i + 1$. Find $\widehat{Q}_i^o(c_i) = \arg,\min_Q\{c_iQ + \mathcal{L}(Q)\}$.

Step 2. Solve problem P with $Q^o = \widehat{Q}_i^o(c_i)$ to decide on the optimal allocation of $\widehat{Q}_i = \min\left\{\widehat{Q}_i^o(c_i), \sum_n U_n\right\}$ to the supplier base.

Step 3. Compute the average unit cost associated with purchasing $\widehat{Q}_i$ units, $c_{i+1} = PC(\widehat{Q}_i)/\widehat{Q}_i$.

Step 4. If the solution converges (i.e., if $\left|\widehat{Q}_i - \widehat{Q}_{i-1}\right| < \epsilon$, where ϵ is a small enough constant) or the algorithm is run for a sufficiently long time, quit with $Q = \widehat{Q}_i$. Otherwise, go to Step 1.

Naturally, the sequential approach described above does not necessarily find the optimal solution. Any approach (such as dynamic programming, DP) that considers the allocation of an additional unit will not guarantee optimality either, as the solution may change drastically by this additional unit. Furthermore, the problem cannot be seen as a special case of a knapsack problem with a non-separable objective function, because the 'knapsack size' (i.e., the total amount to be purchased and allocated to the suppliers) is also a decision variable. Consequently, the problem requires a different solution approach.

Nevertheless, the following DP formulation can be used to solve the integrated problem of finding optimal procurement decisions, including the procurement quantity, with $f_n(x)$ defined as the minimum total cost of

(i) procuring from the partial supplier base $\{n, n+1, \ldots, N\}$ and
(ii) the expected overage and underage of the total quantity purchased from the full supplier base $\{1, \ldots, N\}$,

when x units are already procured from the partial supplier base $\{1, 2, \ldots, n-1\}$.

The manufacturer's problem (MP):

$$\text{for } 0 \leq x \leq \sum_{i=1}^{N} U_i : \quad f_{N+1}(x) = \mathcal{L}(x),$$
$$\text{for } 0 \leq x \leq \sum_{i=1}^{N} U_i : \quad f_n(x) = \min_{y: x \leq y \leq x+U_n} \{C_n(y-x) + f_{n+1}(y)\} \quad \text{for} \quad 2 \leq n \leq N,$$
$$f_1 = \min_{y: 0 \leq y \leq U_1} \{C_1(y) + f_2(y)\}.$$

Theorem 1 *The minimum cost attained by the optimal solution of MP is given by f_1 for any arbitrary order of suppliers numbered from 1 to N.*

Proof Let us number the suppliers from 1 to N. Any order can be used. The Procurement Problem is to find the optimal procurement quantities q_n^* for $n \in \{1, \ldots, N\}$ that minimize the total cost of procuring from the supplier base $\{1, 2, \ldots, N\}$ and the expected overage and underage cost, i.e.,

$$\begin{aligned}
& C_1(q_1^*) + C_2(q_2^*) + \cdots + C_N(q_N^*) + \mathcal{L}(q_1^* + q_2^* + \cdots + q_N^*) \\
&= \min_{\substack{0 \leq q_1 \leq U_1, \\ \ldots, \\ 0 \leq q_N \leq U_N}} \{C_1(q_1) + C_2(q_2) + \cdots + C_N(q_N) + \mathcal{L}(q_1 + q_2 + \cdots + q_N)\} \\
&= \min_{0 \leq q_1 \leq U_1, \ldots, 0 \leq q_N \leq U_N} \{C_1(q_1) + C_2(q_2) + \cdots + C_N(q_N) + f_{N+1}(q_1 + q_2 + \cdots + q_N)\} \\
&= \min_{0 \leq q_1 \leq U_1, \ldots, 0 \leq q_{N-1} \leq U_{N-1}} \Big\{C_1(q_1) + \cdots + C_{N-1}(q_{N-1}) \\
&\quad + \min_{0 \leq q_N \leq U_N} \{C_N(q_N) + f_{N+1}(q_1 + q_2 + \cdots + q_N)\}\Big\} \\
&= \min_{0 \leq q_1 \leq U_1, \ldots, 0 \leq q_{N-2} \leq U_{N-2}} \{C_1(q_1) + \cdots + C_{N-1}(q_{N-1}) + f_N(q_1 + \cdots + q_{N-1})\} \\
&= \min_{0 \leq q_1 \leq U_1, \ldots, 0 \leq q_{N-2} \leq U_{N-2}} \Big\{C_1(q_1) + \cdots + C_{N-2}(q_{N-2}) \\
&\quad + \min_{0 \leq q_{N-1} \leq U_{N-1}} \{C_{N-1}(q_{N-1}) + f_N(q_1 + q_2 + \cdots + q_{N-1})\}\Big\} \\
&= \min_{0 \leq q_1 \leq U_1, \ldots, 0 \leq q_{N-2} \leq U_{N-2}} \{C_1(q_1) + \cdots + C_{N-2}(q_{N-2}) + f_{N-1}(q_1 + \cdots + q_{N-2})\} \\
&\ldots \\
&= \min_{0 \leq q_1 \leq U_1} \{C_1(q_1) + f_2(q_1)\} = f_1.
\end{aligned}$$

Note that the above result does not depend on the ordering of the suppliers due to the commutative property of the addition operator; hence, it does not depend on the initial choice of ordering, and, therefore, the theorem holds for any arbitrary order of suppliers. □

Let $q_n^*(x)$ be such that $x \leq q_n^*(x) \leq x + u_n$ and

$$C_n\left(q_n^*(x)\right) + f_{n+1}\left(x + q_n^*(x)\right) \leq C_n(y - x) + f_{n+1}(y) \quad \forall y : x \leq y \leq x + u_n,$$

for any given value of x. Then, the optimal quantity procured from supplier n, τ_n, is given by

$$\tau_1 = q_1^*(0), \tau_n = q_n^*\left(\sum_{i=1}^{n} \tau_i\right) \text{ for } 2 \leq n \leq N.$$

The total optimal procurement quantity is given by $Q^* = \sum_{i=1}^{N} \tau_i$. The computational complexity of this DP is $O\big(N\big(\sum_n U_n\big)\max_n(U_n)\big)$.

We conducted a numerical study to investigate (i) the effect of problem parameters on the optimal solution (Sects. 8.2.1 and 8.2.2) and (ii) the performance of the sequential approach (Sect. 8.2.3). We considered the following setting: The demand has a Gamma distribution with coefficient of variation (CV) values of 0.5, 1, 1.5, and with expected values, *E[W]*, of 20, 40, 50, and 60. Demand is assumed to be discrete in this section for ease of exposition. The cost parameters are $h = 1$, $b = 2, 5, 10, 50$, and 200. We consider three sets of suppliers. In the first set (Supplier Base 1), there are $N = 5$ alternative suppliers $(n = 1, 2, \ldots, 5)$ with capacities U_n = 40, 20, 20, 10, and 10, respectively. There exists a fixed-cost component of ordering from supplier n, with $K_n = 40, 20, 20, 10$, and 10, respectively, and a linear unit variable cost component of c_n, in which $c_1 \in \{1.5, 2, 2.5\}$, c_2 and $c_3 \in \{2, 2.5, 3\}$, c_4 and $c_5 \in \{2.5, 3, 3.5\}$. This set resembles a situation in which the supplier base consists of a variety of suppliers, in terms of cost and capacity. In the second set (Supplier Base 2), there are also $N = 5$ alternative suppliers, but their capacities are $U_n = 60, 10, 10, 10$, and 10, respectively. We set the fixed cost of ordering from supplier n as $K_n = 60, 10, 10, 10$, and 10, respectively, and we set a linear unit variable cost component of c_n, as $c_1 \in \{1.0, 1.5, 2.0\}$, and c_2 to $c_5 \in \{2.5, 3, 3.5\}$. This set resembles a situation in which there is one dominant supplier in the supply base, and the rest are relatively smaller suppliers. The third set (Supplier Base 3) also consists of $N = 5$ alternative suppliers, but their capacities are $U_n = 24, 22, 20, 18$, and 16, respectively. We set the fixed cost of ordering from supplier n as $K_n = 24, 22, 20, 18$, and 16, respectively, and a linear unit variable cost component of c_n, as $c_1 \in \{1.8, 2.3, 2.8\}$, $c_2 \in \{1.9, 2.6, 2.9\}$, $c_3 \in \{2.0, 2.5, 3.0\}$, $c_4 \in \{2.1, 2.6, 3.1\}$, and $c_5 \in \{2.2, 2.7, 3.2\}$. This set resembles a situation in which there is no dominant supplier, and all suppliers are comparable in capacity.

Table 8.2 The optimal procurement decision at different coefficients of demand variation and underage costs

	$b=2$	$b=5$	$b=10$	$b=50$	$b=200$
CV = 0.5	$(0,0,0,0,0)$	$(40,0,0,0,0)$	$(40,0,0,0,0)$	$(40,20,17,0,0)$	$(40,20,20,10,0)$
CV = 1.0	$(0,0,0,0,0)$	$(0,20,0,0,0)$	$(40,0,0,0,0)$	$(40,20,20,10,0)$	$(40,20,20,10,10)$
CV = 1.5	$(0,0,0,0,0)$	$(0,0,0,0,0)$	$(40,0,0,0,0)$	$(40,20,20,10,10)$	$(40,20,20,10,10)$

Supplier Base 1; $E[W]=40$; and $c_n = 1.5, 2, 2, 3$, and 3, for $n = 1, \ldots, 5$

8.2.1 Effects of Demand Variability and Cost Parameters

The following insight that simple inventory/production models generate holds for the procurement problem to some extent: As the unit underage cost increases (while keeping all other problem parameters constant), the total quantity procured from the suppliers and the total expected costs of the operation increase. As any cost component of a supplier increases, the supplier is preferred less by the buyer, and the total procurement quantity, if any, from that supplier decreases. The optimal total procurement quantity does not necessarily increase as the variability of demand increases (see Table 8.2), because the risk of being left with unsold goods (as in obsolescence) outweighs the risk of goodwill loss, due to relatively high procurement and overage costs.

We also observe that the optimal solution might be extremely sensitive to cost parameters. For example, when $\mathrm{CV}=1.0$, $b=5$, $N=3$, $U_n=40, 20, 10$, $K_n = 40, 20, 10$, and $c_n = 1.5, 2.5, 2.5$, for $n = 1, 2, 3$, respectively, the optimal solution is $(37, 0, 0)$. When we keep all parameters the same, except for $c_1 = 2$, instead of 1.5, the optimal solution becomes $(0, 0, 10)$, which represents not only a 73 % decrease in total procurement quantity, but also a completely different supplier selection. This example shows that the optimal solution of a particular situation could significantly change, even when a single parameter changes, indicating a lack of robustness, which emphasizes the importance of having a methodology appropriate for finding the optimal solution.

8.2.2 Effects of Flexibility

In this section, we analyze the impact of flexibility on optimal procurement decisions. We call one problem environment 'more flexible' than another when there is at least one more procurement option to choose from. In our numerical tests, we frequently observe that the total procurement quantity does not decrease as the problem environment becomes more flexible. Nevertheless, our numerical experiments reveal that a more flexible environment may also lead to lower procurement quantities. Such a situation is observed when a more appealing (e.g., cheaper per unit, when the order size is sufficiently high) procurement alternative is introduced to the supplier base, and it is not necessary to place a high order size, to benefit from economies of scale in the former situation, by utilizing this new supplier. An example of this situation can be illustrated by the following instance: Supplier Base 1, $E[W] = 40$, $\mathrm{CV} = 0.5$, $b = 5$, $c_n = 2.5, 3, 3, 2.5$, and 2.5, for $n = 1, \ldots, 5$, respectively. Let us first set $U_3 = U_5 = 0$, i.e., only suppliers 1, 2, and 4 are available with $U_1 = 40, U_2 = 20$, and $U_4 = 10$. In this case, the optimal solution is $(34, 0, 0, 0, 0)$. When we make this system more flexible by letting $U_3 = 20$, and $U_5 = 10$, the optimal solution becomes $(0, 0, 0, 10, 10)$, decreasing the total procurement by 41 %. In the former situation, the buyer does not prefer to procure 20 units (as in the latter case), because Supplier 1 is short in capacity, and Supplier 2 is a more expensive option. The fixed cost of Supplier 1 leads to the procurement of a larger quantity in the optimal solution. In the latter situation, the introduction of Supplier 5, a cheaper option, makes it unnecessary to utilize Supplier 1 with its high fixed cost; 20 units turn out to be optimal when the trade-off between the underage and fixed costs are resolved. This phenomenon is observed in several more problem instances with similar conditions, also for Supplier Base 2 and 3. On the other hand, this is attributed to the existence of suppliers with diverse cost and capacity structures, e.g., when there is a dominant supplier. When we decreased this diversity in our numerical tests by trimming the cost differences among the suppliers in any particular supplier base, we consistently observed a decrease in the number of cases in which this phenomenon is observed. Obviously, in the limit when all suppliers are identical, increasing flexibility does not lead to a decrease in total procurement quantity.

8.2.3 Value of the Integrated Approach

Finally, we compare the optimal solution of the integrated approach with the solution found by the sequential approach as presented in Sect. 8.2. The average cost deviation percentages relative to the optimal solution over all the problems in our test bed are presented in Fig. 8.1.

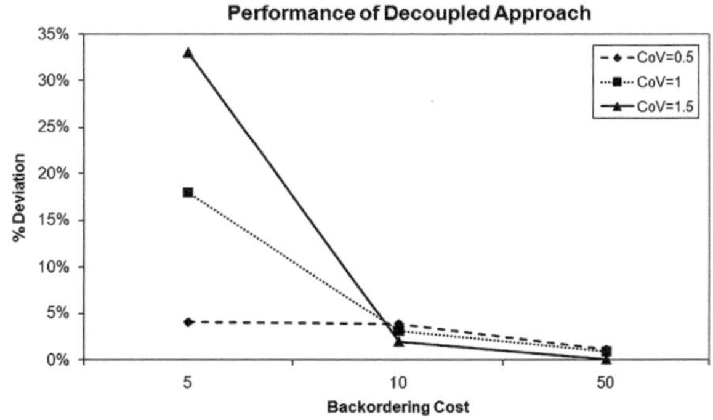

Fig. 8.1 Percent of cost deviation due to sequential approach

The sequential approach performs well, when the underage cost is either extremely low or is the dominating cost factor: the sequential approach yields the optimal solution when $b = 2$ and the average cost deviation is 0.34 % for $b = 200$, in our test bed. This is because when b is as low as 2 in our test bed, the optimal policy is trivially to always backorder; when b is high, procurement takes place in large quantities, sometimes consuming the full available capacity, which is the other extreme trivial solution. Nevertheless, when the underage cost is neither dominating nor insignificant, the value of the integrated approach over the sequential approach appears to be significant. The average cost deviation over all cases considered is 12.18 and 4.67 % when $b = 5$ and 10, respectively; the maximum is 85.81 %, which also demonstrates the importance and non-triviality of finding the optimal solution.

8.3 Supplier's Problem: How to Cast a Counterspell?

In this section, we take the suppliers' point of view into consideration. Consider a particular supplier (referred to as 'the supplier' from now on) who intends to earn the manufacturer's business, preferably by forming a channel between himself and the manufacturer. The supplier—who might be a new entrant to the market—intends to respond to the RFQ announced by the manufacturer. He either needs to install new capacity or spare a portion of his existing capacity for the manufacture of the item. What the supplier must determine are the optimal capacity to dedicate and the optimal price to quote to the manufacturer, under the existence of other suppliers (referred to as the 'alternative suppliers' in the rest of the text). After receiving all quotations, the manufacturer will determine her optimal course of action by using the methodology explained in Sect. 8.2.

The supplier would benefit from the price and capacity information of the alternative suppliers in order to make a better decision about the capacity to dedicate and the price to quote to the manufacturer, should the information be collected one way or another. Such information might be available to the supplier if (i) the supplier has enough experience in the market, e.g., through the subcontractors that he has been collaborating with, (ii) the majority of the alternative suppliers are members of organized industrial associations or zones, where their association puts additional marketing effort by disclosing relevant information to interested parties, (iii) there exists a business-to-business establishment, e.g., an e-business portal with suppliers' posted price and capacity information (see, e.g., Agrali et al. 2008, for the case of an auction-based logistics market), or an online auction (see, e.g., Chen et al. 2005) where the information is made available to the other bidders, and the like. Note that in some cases such as sourcing from overseas suppliers, transportation cost might constitute an important portion of the procurement cost, which facilitates collecting necessary information on the cost structure. If the supplier has information on the manufacturer's demand distribution or he can anticipate it, he could use the methodology presented in Sect. 8.2 to predict how the manufacturer would operate. The question that we address in this section is how the supplier can make use of this information to form a list of price quotations at various quantities that will result in the manufacturer procuring the quantity that maximizes the supplier's profit. If the supplier could find such a price and capacity pair, then he would eliminate the uncertainty as to the capacity he should dedicate to this manufacturer.

Prior to the quotation of the supplier, the manufacturer has a certain course of action. However, any capacity and price quotation offered by the supplier might change the manufacturer's decision considerably. As noted in Sect. 8.2.1, this problem is very sensitive to the problem parameters; the effect of a change in even one of the parameters or an increase in the number of available suppliers cannot be easily anticipated without solving the problem under the new settings to optimality. Therefore, even if the supplier has all the necessary information, it is not straightforward to derive insights and to set a price and capacity pair without a methodology to find the optimal solution. The supplier needs to solve the following optimization problem, where we index the supplier as 1, and the alternative suppliers from 2 to N without loss of generality.

The supplier's problem (SP):

$$\begin{aligned} \max_{p(Q^s)\geq 0, 0<Q^s\leq U_1} \quad & Z(p(Q^s), Q^s) = C_1(Q^s) - K_1(Q^s) - A(Q^s) \\ \text{s.t.} \quad & C_1(Q^s) + f_2(Q^s) \leq C_1(y) + f_2(y) \quad \forall y \leq U_1 \\ & Q^s, y : \text{integer} \end{aligned} \tag{8.1}$$

where

Q^s	Quantity quoted by the supplier
$p(Q^s)$	Average price per unit corresponding to selling Q^s units
$K_1(Q^s)$	Costs associated with purchasing Q^s units from the supplier that are not accrued by the supplier
$Z(p(Q^s), Q^s)$	Total profit of the supplier
$A(Q^s)$	Total cost of dedicating Q^s units of capacity to the manufacturer
$C_1(Q^s)$	$Q^s p(Q^s) + K_1(Q^s)$
$f_2(Q^s)$	The minimum cost of purchasing from alternative suppliers. If Q^s units are purchased from the supplier

Recall that $f_2(Q^s)$—which is the minimum cost of purchasing from all of the alternative suppliers—does not depend on the ordering of the suppliers, due to Theorem 1. Hence, the alternative suppliers may be ordered arbitrarily from 2 to N, where the solution does not depend on which supplier is indexed as number 2.

The objective function of SP is to maximize the profit generated by the supplier when the manufacturer purchases Q^s units with a cost of $C_1(Q^s)$, resulting is an average price of $p(Q^s)$ per unit, accrued by the supplier (possibly as a result of a nonlinear cost scheme quoted by the supplier). The cost $C_1(Q^s)$ also includes all costs associated with purchasing Q^s units from the supplier that are not accrued by the supplier, such as the shipping costs charged by a logistics service provider. In the constraint set, the expression on the left-hand side is the total cost of the manufacturer's optimal purchasing strategy when the supplier quotes Q^s units at an average price of $p(Q^s)$ per unit, whereas the right-hand side is the manufacturer's total cost associated with procuring any quantity less than U_1 from the supplier and the rest from the alternative suppliers. This constraint set ensures that the price quoted for each Q^s value makes it economical for the manufacturer to procure Q^s units in full from the supplier with a cost of $C_1(Q^s)$. Note that SP is a nonlinear programming model as the functions $A(Q^s)$, $C_1(Q^s)$, and $f_2(Q^s)$ can have any functional form. Nevertheless, we devise an algorithm to find the optimal solution by inspection.

For a given value of Q^s, $p(Q^s)$ attains the largest possible value, since we have a maximization problem. We first note that the constraint (8.1) at $y = 0$ provides an upper bound on $p(Q^s)$ because the manufacturer would procure only from the alternative suppliers for any price quotation above $p(Q^s)$. Since $C_1(0) = 0$, this upper bound turns out to be $p(Q^s) \leq \frac{f_2(0) - f_2(Q^s)}{Q^s}$. Repeating this for all $0 < Q^s \leq U_1$ generates a list of price quotations at each possible Q^s such that the manufacturer is indifferent between procuring Q^s units at a price of $p(Q^s)$ from the supplier and procuring Q^s units elsewhere. This means that the constraint set (8.1) is equivalent to $C_1(Q^s) + f_2(Q^s) \leq f_2(0) \quad \forall 0 < Q^s \leq U_1$, which decreases the complexity of the problem.

While SP generates a list of price quotations for all $0 < Q^s \leq U_1$, the supplier would not be interested in $(Q^s, p(Q^s))$ pairs with $Z(p(Q^s), Q^s) \leq 0$. Hence, the list consists of the $(Q^s, p(Q^s))$ pairs with positive profit. The supplier needs to give an incentive to the manufacturer to make sure that $Q^{s*} = Q^s$ that maximizes $Z(p(Q^s), Q^s)$ is procured by quoting a price of $p^*(Q^{s*}) = p(Q^{s*}) - \varepsilon$ for Q^{s*}, with $\varepsilon > 0$. We note that $Z(p^*(Q^{s*}), Q^{s*})$ is the maximum benefit that can be generated by the business channel between the supplier and the manufacturer. The supplier enjoys all of this benefit but the incentive, where the incentive ensures that the manufacturer is also better off compared to the situation without this business channel, resulting in channel coordination.

Property 2 *A list is optimal if it includes* $(Q^{s*}, p^*(Q^{s*}))$ *and* $(\bar{Q}^s, p(\bar{Q}^s))$ *such that* $p(\bar{Q}^s) \geq p(Q^s)$ *for all* $Q^s \neq Q^{s*}$.

In what follows, we provide an algorithm that can be used to generate a profitable quotation list, based on SP.

Step 0. Number the alternative suppliers starting from 2 and find $f_2(Q^s)$ by solving MP for all $0 \leq Q^s \leq U_1$.

Step 1. For each value of Q^s such that $0 \leq Q^s \leq U_1$, $p(Q^s) = (f_2(0) - f_2(Q^s))/Q^s$.

Step 2. Let $Q^{s*} = \arg\max_{Q^s} Z(p(Q^s), Q^s)$ and $p^*(Q^{s*}) = p(Q^{s*}) - \varepsilon$, such that $p^*(Q^{s*}) > 0$ and $Z(p^*(Q^{s*}), Q^{s*}) > 0$. If no such $(p^*(Q^{s*}), Q^{s*})$ exists, quit the algorithm as there is no profitable quotation list.

Step 3. An optimal quotation list consists of $(Q^{s*}, p^*(Q^{s*}))$ and $(Q^s, p(Q^s))$ for all $0 < Q^s \leq U_1$ such that $Q^s \neq Q^{s*}$, $p(Q^s) > 0$ and $Z(p(Q^s), Q^s) > 0$.

Step 0 and Step 1 take $O\left(N\left(\sum_n U_n\right)\max_n(U_n)\right)$ and $O(U_1)$ computational time, respectively, and Step 2 can already be computed within the effort required in Step 1. Therefore, the computational complexity of this algorithm is $O\left(N\left(\sum_n U_n\right)\right.$ $\left.\max_n(U_n)\right)$, i.e., it does not add to the complexity of MP.

Although for any non-speculative cost structure $f_2(Q^s)$ is non-increasing in Q^s, we note that $p(Q^s)$ is not necessarily monotonic in Q^s. Figure 8.2 depicts an example under the parameter setting introduced in Sect. 8.2 with $b = 10$, CV $= 1.5$, and $c_n = 1.5, 2, 2, 2.5, 2.5$ for $n = 1, \ldots, 5$, respectively (the dotted line on the figure). The optimal $(Q^{s*}, p^*(Q^{s*}))$ is also encircled in Fig. 8.2, which is not even a local optima of the Q^s versus $p(Q^s)$ graph. While the Q^s versus $p(Q^s)$ graph may yield any form, it would be unconventional and complicated to quote such a non-monotone pricing scheme. To that end, the supplier can adopt a more practical scheme, such as quantity discounts, as long as it is in line with Property 2. Note that such a scheme would require the supplier to apply artificial mark-ups to optimal prices. We also depict an example pricing scheme with quantity discounts after the artificial mark-up in Fig. 8.2 (the solid line). A possible disadvantage of quoting elevated prices in practice is the prospective loss of goodwill of the manufacturer. Therefore, a remedy would be to apply a constant unit-price scheme (or, 'linear' scheme), which is observed frequently in practice. Furthermore, the manufacturer might specifically require a linear scheme. Nevertheless, which constant unit price must be quoted is not a trivial decision and requires further analysis. Quoting $p^*(Q^{s*})$ is not necessarily optimal, and moreover, it violates Property 2 unless $p^*(Q^{s*}) = \max_{Q^s} p(Q^s)$. Therefore, in the remainder of this section, we consider the 'special case' of linear price quotations between the supplier and the manufacturer.

If the manufacturer requires a linear pricing scheme from the supplier, then the supplier's problem becomes the following:

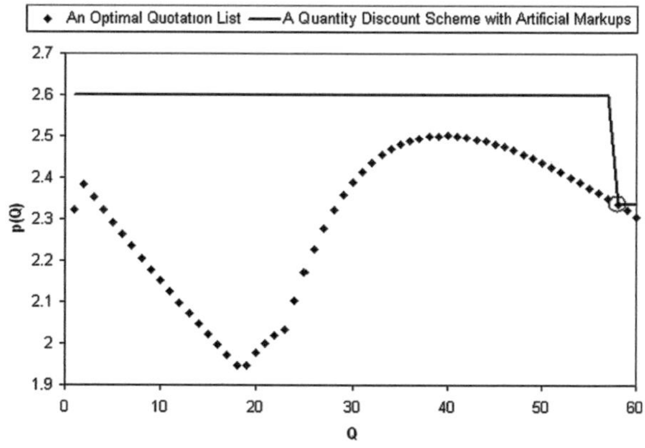

Fig. 8.2 An optimal pricing scheme example

$$
\begin{array}{lll}
(\mathrm{SP^L}): & \max\limits_{p \geq 0, 0 < Q^s \leq U_1} & Z(p, Q^s) = pQ^s - A(Q^s) \\
\text{s.t.} & & pQ^s + K_1(Q^s) + f_2(Q^s) \leq py + K_1(y) + f_2(y) \quad \forall y \leq U_1 \\
& & Q^s, y: \text{integer}
\end{array}
$$

For any $0 < Q^s \leq U_1$, we have the following relations from the constraint set:

$$
p \quad \leq \quad (f_2(y) - f_2(Q^s) + K_1(y) - K_1(Q^s))/(Q^s - y) \quad \text{for all } y \leq U_1 \tag{8.2}
$$

Let the optimal price to quote that would result in ordering Q^s units from the supplier be $\bar{p}(Q^s)$. The maximum unit price that will not violate (8.2) is given by $\bar{p}(Q^s) = \min_y (f_2(y) - f_2(Q^s) + K_1(y) - K_1(Q^s))/(Q^s - y)$. In what follows we provide an algorithm that can be used to generate a profitable quotation list, based on $\mathrm{SP^L}$:

Step 0. Number the alternative suppliers starting from 2 and find $f_2(Q^s)$ by solving MP for all $0 \leq Q^s \leq U_1$.
Step 1. For each value of Q^s such that $0 \leq Q^s \leq U_1$, $\bar{p}(Q^s) = \min_{y: 0 \leq y \leq U_1} (f_2(y) - f_2(Q^s) + K_1(y) - K_1(Q^s))/(Q^s - y)$.
Step 2. Let $Q^{s*} = \arg\max_{Q^s} Z(\bar{p}(Q^s), Q^s)$ and $p^* = \bar{p}(Q^{s*})$ such that $p^* > 0$ and $Z(p^*, Q^{s*}) > 0$. If no such (p^*, Q^{s*}) exists, quit the algorithm as there is no profitable quotation.
Step 3. The optimal unit price is to quote available capacity U_1 at a unit price of p^*.

By solving $\mathrm{SP^L}$, the supplier finds the optimal quantity Q^{s*} that will be ordered by the manufacturer from the quoted capacity of U_1, and the corresponding unit price p^* that will maximize his profit. If the algorithm generates a non-empty quotation list, then the supplier will be in business. In this case, the manufacturer is also better off and benefits due to the presence of the supplier.

In the following discussion, we examine the impact of problem parameters on operating characteristics. As a numerical test bed, we use the parameter set introduced above, and in addition we let $A(Q^s) = 1.5Q^s$ and include $b = 100$. For this discussion, let Π^s denote the benefit (i.e., the profit) of the supplier, Π^m the benefit of the manufacturer, and $\Pi = \Pi^s + \Pi^m$ the total benefit of the system due to the presence of the supplier. If the supplier decides not to engage in business due to a non-positive profit, then the benefits are zero. We first investigate the impact of the demand variability on the supplier's and manufacturer's benefits under different backordering costs (see Fig. 8.3).

For low values of the backordering cost ($b = 5$ or 10 in Fig. 8.3), we observe that the benefit to the supplier decreases as the demand variability increases. This is because the manufacturer prefers to decrease[1] the total procurement amount from the market (see Table 8.3), cf. Section 8.2.1. For larger values of b, the

[1]In this discussion, we use the term 'decreasing' ('increasing') in the weak sense, to mean 'non-increasing' ('non-decreasing').

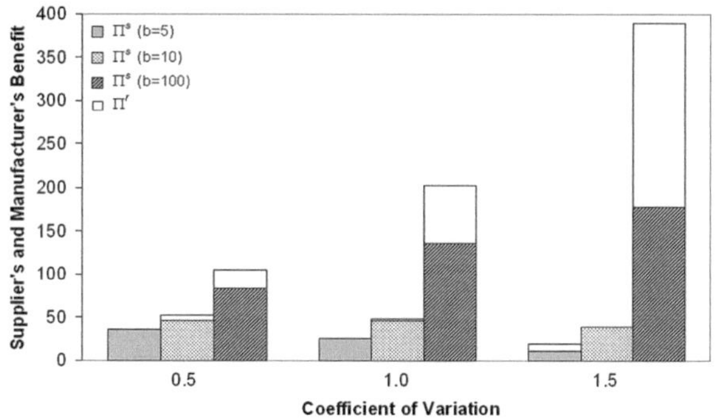

Fig. 8.3 Impact of demand variability on the benefits to the supplier and the manufacturer

Table 8.3 Optimal procurement quantities and the unit price

	Q^{s*}			Q_m			p^*		
CV	$b = 5$	$b = 10$	$b = 100$	$b = 5$	$b = 10$	$b = 100$	$b = 5$	$b = 10$	$b = 100$
0.5	33	47	84	33	47	84	2.59	2.5	2.49
1	17	47	90	17	47	130	3.02	2.48	3
1.5	10	39	89	10	39	169	2.64	2.5	3.5

Q_m : Capacity procured by the manufacturer from all suppliers

manufacturer's total procurement quantity increases in demand variability, which leads to an increase in the benefit to the supplier.

The optimal prices quoted by the supplier under different demand variations and backordering costs are also shown in Table 8.3. The behavior of the optimal price strongly depends on the problem parameters and, in general, there is no monotonicity. When $b = 100$, we observe that p^* increases as CV increases, even though Q^{s*} remains about the same. Recall that as CV increases, the manufacturer is willing to procure more capacity under high backordering costs. This makes the supplier's capacity more valuable and gives him an opportunity to elevate his prices. To be more specific, we explain the rationale behind this opportunistic behavior as follows: In this problem instance, a total of 100 units of capacity are available from alternative suppliers, and the maximum capacity that the supplier can quote is also 100 units. When CV = 0.5, the supplier competes with all alternative suppliers for the existing range of capacity that would be procured by the manufacturer and quotes a price of 2.49, which beats all alternative suppliers. When CV = 1, the manufacturer is willing to procure more than 100 units in total. As 40 units are procured from the 'cheapest' alternative supplier, the supplier now competes with the remaining 'relatively more expensive' ones that have a total capacity of 60 and hence is able to increase his quoted price while achieving higher sales (90 units rather than 84). When CV = 1.5, the supplier competes with even more expensive suppliers and hence increases the quoted price.

A similar but reverse effect is observed for $b = 5$. When CV is increased from 1 to 1.5, the manufacturer prefers to procure less this time, since the backordering

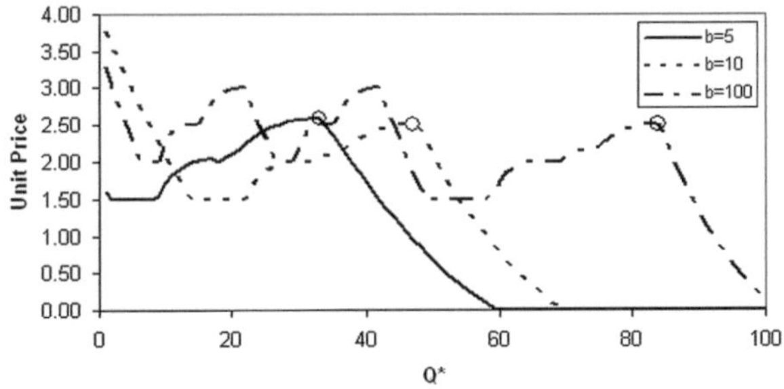

Fig. 8.4 Quantity procured from the supplier's quoted capacity versus unit price, when CV = 0.5

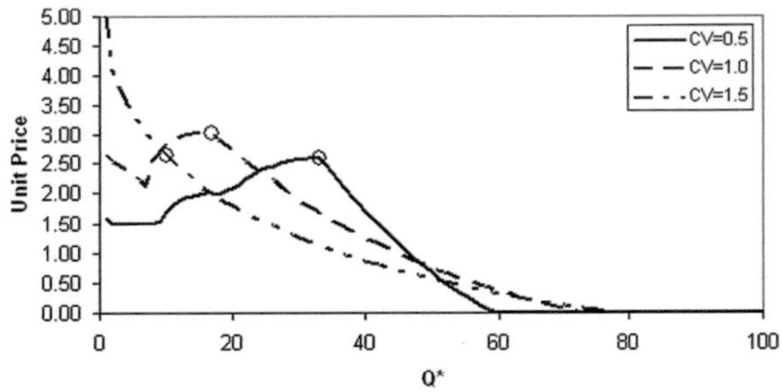

Fig. 8.5 The supplier's quoted capacity versus unit price

cost is relatively low. Hence, the supplier needs to compete for a smaller market size with 'relatively cheaper' suppliers, which forces him to decrease the quoted prices.

When $b = 10$, the manufacturer normally prefers to procure less as CV increases from 0.5 to 1 (all other parameters are kept constant). However in this case, the supplier slightly decreases his price (from 2.5 to 2.48) by forcing the manufacturer to procure the same quantity (47). Had the manufacturer procured 46 units at a price of 2.5, the supplier's profit would have been 115, which is less than the profit he makes (116.56) by selling 47 units at a price of 2.48.

As illustrated above, the optimal prices are determined according to the particular interactions of the problem parameters and there is no monotonic behavior. Figure 8.4 (Fig. 8.5) depicts the optimal quantity procured from the supplier versus the unit price quoted, for different backordering costs when CV = 0.5 (coefficient of variation values when $b = 5$). The optimal procurement quantity and unit-price pair are shown with a circle. In both figures, we plot the graphs for all unit prices in the feasible range, irrespective of profitability. As the unit cost for the supplier is 1.5, quoting any price less than 1.5 would not be rational in the short run; nevertheless, the supplier might prefer to operate with negative profits in return for capturing a large portion of the market and garnering a strategic benefit in the long run.

Finally, we elaborate on the benefit of channel coordination under the linear pricing scheme as modeled by SP^L, making use of a numerical example with a Poisson demand and $b = 100$. In this case, the optimal course of action for the supplier is to quote a unit price of 2.5, which results in a procurement quantity of 52 units with $\Pi^s = 52$, $\Pi^m = 20.84$, and $\Pi = 72.84$. This is the only operational point that would be materialized without any further coordination effort. However, the manufacturer's benefit would have been maximized if she had requested 39 units from the supplier, which would dictate the supplier to quote a unit price of 2 according to SP^L. In this case, $\Pi^s = 19.50$, $\Pi^m = 68.15$, and $\Pi = 87.65$. Nevertheless, neither of these two operating points coordinate the channel. The maximum benefit of the channel is attained when the supplier quotes a unit price of 2.14, resulting in a procurement quantity of 41 units, with $\Pi^s = 26.28$, $\Pi^m = 62.24$, and $\Pi = 88.52$. A particular mechanism in the form of a tailored contract is necessary to ensure that both parties are better off and this channel-coordinating point is attained. Hence, the maximum channel profit of 88.52, which stands for an additional benefit of 15.68 compared to the situation without coordination, could be shared between the parties in such a way that the supplier's profit exceeds 52 and the manufacturer's benefit exceeds 20.84.

8.4 Manufacturer's in-House Production Capacity Problem: Cast Your Own Spell

In this section, we switch back to the manufacturer's point of view, with the consideration that she might allocate some in-house production capacity to produce the item if she has (the ability to acquire) the technology to do so. This might be desirable for the manufacturer not only because of cost advantages, but also due to the strategic decision of being less dependent on suppliers. Moreover, the solution to MP is highly sensitive to relatively small changes in problem parameters, as discussed in Sect. 8.2.1, and the manufacturer might need to build or allocate some in-house capacity as a remedy. Assuming that such concerns can be translated into financial terms (i.e., updating the quotations accordingly to incorporate them), we take the cost perspective into account in what follows.

As the quotations of prospective suppliers are available to the manufacturer, she may be better off manufacturing (part of) the items in-house, depending on the quotations and the cost of allocating her own manufacturing capability or acquiring this capability. Therefore, the manufacturer makes the in-house production versus outsourcing decision, where combining the two is also an option. The methodology we introduced in Sect. 8.2 can be used as the key facilitator to that end. We note that it does not suffice to simply use 'in-house production option' as an alternative supplier in that methodology, because the capacity to allocate is also a decision variable now. Nevertheless, the manufacturer can determine her optimal course of action in terms of best in-house capacity allocation versus outsourcing strategy as follows:

Let the cost of acquiring/allocating in-house production capability for manufacturing Q^{ih} items be $A(Q^{ih})$, which may assume any form. Then, the manufacturer's in-house production capacity problem (MCP) can be modeled as follows:

$$\min_{0 \le Q^{ih} \le U_0} A(Q^{ih}) + f_2(Q^{ih})$$

where U_0 is the maximum in-house capacity that can be allocated. Note that this model considers all possible outsourcing options in combination with in-house production in one shot. The solution complexity is the same as that of MP, i.e., $O\left(N\left(\sum_n U_n\right)\max_n(U_n)\right)$. That is, once the cost structure of allocating in-house capability is known, there is no additional complexity required for solving MCP.

MCP shows a similarity to SP, as the capacity to be quoted is also a decision variable in SP. Nevertheless, the objective of the supplier is to maximize his profit, whereas that of the manufacturer is to minimize her costs. The maximum benefit that can be generated by introducing a 'new source' of capacity (i.e., the supplier's capacity in Sect. 8.3 and the in-house option here) to the system is the same in both models. Hence, if $A(\cdot)$ is the same in those two models and the quotation list of the supplier is determined by the solution of SP as proposed with the algorithm provided in Sect. 8.3 with $\varepsilon = 0$, the difference between those two cases rests on who collects the benefit, and the total production remains the same. Nevertheless, the total benefit generated with different quotation structures (as in SP^{L}) might be less than that with MCP, which might encourage the manufacturer to produce in-house and eventually avoid double marginalization. Similarly, the total production quantities with MCP and with different quotation structures (as in SP^{L}) are also not necessarily the same.

8.5 Conclusions

In this paper, we consider the sourcing decisions of a manufacturer in fashion industry from three perspectives: (i) Supplier selection problem of the manufacturer where she determines which supplier(s) to utilize and to what extent, (ii) Capacity and price quotation problem of a supplier, (iii) In-house versus outsourcing decision of the manufacturer. We allow for stochastic demand and capacitated production facilities. Our modeling approach is capable of handling sourcing problems in a wide range of environments, as we do not impose restrictions on the relevant cost components. The procurement problem and its several variations are proven to be NP-hard in literature; however, we develop a dynamic programming model with a state definition, which makes the solution algorithm pseudo-polynomial. We achieve this by proving that the order of the sources is irrelevant for the optimal solution. Our main model is the basis for solving all three subproblems posed.

We derive the following managerial insights through numerical studies:

- An increase in the availability of sourcing options (a more flexible system) may lead to a decrease in the total quantity procured, when there are suppliers with diverse cost and capacity structures, e.g., when there is a dominant supplier.

- The optimal solution to the sourcing problem is not necessarily robust, as a change in even a mere cost parameter might completely change the optimal course of action. In case robustness is sought (for reasons such as ensuring product uniformity or decreasing administrative costs of procurement), strategic partnership, vertical integration, or making instead of buying are some possible means of eliminating or reducing such parameter dependence.
- As it is also common to newsvendor models, the total quantity procured by the manufacturer does not necessarily increase as variability of demand increases. For relatively low service level requirements, the total quantity procured decreases as the variability of the demand increases; whereas a reverse effect is observed otherwise.
- There is significant value in integrating the decisions as to the supplier selection and the procurement quantity, particularly for moderate service level requirements.
- The entrance of a new supplier to the market can form a business channel between the supplier and the manufacturer, which brings a nonnegative benefit to both parties (in terms of decreased sourcing costs for the manufacturer and profit for the supplier). The party that reaps the maximum benefit that can be generated is the supplier, as long as he has the liberty of setting a quotation list in any form, such as non-monotonically quoted prices. As such a quotation list might be impractical, the supplier may be forced to adopt a particular pricing scheme such as a constant unit price. However, in that case, the generated channel benefit might be limited and is shared by the supplier and the manufacturer. Consequently, the supplier and the manufacturer need to collaborate and tailor a contract in order to ensure that the channel is coordinated and both parties are better off. Traditional policies proposed for channel coordination such as quantity discounts, buy back policies do not necessarily 'do the trick' for coordinating the channel.

We also note that the methodology that we propose can be used repeatedly by relevant decision makers. For example, once the supplier solves SP and offers a quotation list, the manufacturer solves MP (or MCP if there is in-house manufacturing capability and desire) and contacts (some of the) suppliers if necessary for a reverse auction with the motivation of driving the prices down. In that case, if the quotations are disclosed (possibly the supplier identities being censored), any supplier might (re-) solve SP with updated information and offers a new quotation list, provided that it is profitable to do so. Note that multiple suppliers cannot approach this problem in a game theoretical framework, as the *cost* structures of the suppliers—unlike the price–would not be available to each other.

Our work can be extended to include some other relevant elements such as multiple-period decision making or supply disruptions. Furthermore, other possible extensions include explicit treatment of the non-biddable price factors such as delivery punctuality, the quality of items procured, and strategic partnership concerns that we assumed to have been implicitly reflected on the procurement costs.

Multiple criteria analysis taking these factors into account could be another interesting extension.

References

Agrali S, Tan B, Karaesmen F (2008) Modeling and analysis of an auction-based logistics market. Eur J Oper Res 191:272–294

Aissaouia N, Haouaria M, Hassini E (2007) Supplier selection and order lot sizing modeling: a review. Comput Oper Res 34:3516–3540

Alp O, Tan T (2008) Tactical capacity management under capacity flexibility. IIE Trans 40:221–237

Alp O, Huh T, Tan T (2014) Inventory Control with Multiple Set-Up Costs. Manuf Serv Oper Manage 16:89–103

Andrade-Pineda JL, Canca D, Gonzalez-R PL (2015) On modelling non-linear quantity discounts in a supplier selection problem by mixed linear integer optimization. Ann Oper Res. doi:10.1007/s10479-015-1941-2

Ariba (2014) http://www.ariba.com

Awasthi A, Chauhan SK, Proth J-M (2009) Supplier selection problem for a single manufacturing unit under stochastic demand. Int J Prod Econ 117:229–233

Ayhan MB, Kilic HS (2015) A two stage approach for supplier selection problem in multi-item/multi-supplier environment with quantity discounts. Comput Ind Eng 85:1–12

Burke GJ, Carillo J, Vakharia AJ (2008a) Heuristics for sourcing from multiple suppliers with alternative quantity discounts. Eur J Oper Res 186:317–329

Burke GJ, Geunes J, Romeijn HE, Vakharia AJ (2008b) Allocating procurement to capacitated suppliers with concave quantity discounts. Oper Res Lett 36:103–109

Chauhan SS, Proth J-M (2003) The concave cost supply problem. Eur J Oper Res 148:374–383

Chauhan SS, Eremeev A, Kolokolov A, Servakh V (2005) Concave cost supply management for single manufacturing unit. Pardalos PM, Hearn DW (eds) Supply chain optimisation product/process design, facility location and flow control. Springer, US, pp 167–174

Chen RR, Roundy RO, Zhang RQ, Janakiraman G (2005) Efficient auction mechanisms for supply chain procurement. Manage Sci 51:467–482

Davis EW, Page KL (1991) Emerson electric company ACP division: the fan subpack sourcing decision. Darden School of Business Case, University of Virgina, Charlottesville, VA, USA

Elmaghraby WJ (2000) Supply contract competition and sourcing policies. Manuf Serv Oper Manage 2:350–371

Fiatech (2014) http://fiatech.org

Ford (2014) https://fsp.portal.covisint.com/web/portal/home

Foster Wheeler (2014) http://eprocure.fwc.com

Haberl J (1999) Fixed-charge continuous knapsack problems and pseudogreedy solutions. Math Prog 85:617–642

Hazra J, Mahadevan B (2009) A procurement model using capacity reservation. Eur J Oper Res 193:303–316

Hilton (2014) http://www.hiltonworldwide.com/development/performance-advantage/supply-management/e-procurement/

Ho W, Xu X, Dey PK (2010) Multi-criteria decision making approaches for supplier evaluation and selection: a literature review. Eur J Oper Res 202:16–24

Hosseininasab A, Ahmadi A (2015) Selecting a supplier portfolio with value, development, and risk consideration. Eur J Oper Res 245:146–156

Jia P, Govindan K, Choi T-M, Rajendran S (2015) Supplier selection problems in fashion business operations with sustainability considerations. Sustainability 7:1603–1619

Kaplan S, Sawhney M (2000) E-Hubs: the new B2B marketplaces. Harvard Bus Rev 97–103
Kawtummachai R, Hop NV (2005) Order allocation in a multiple-supplier environment. Int J Prod Econ 93–94:231–238
Kheljani JG, Ghodsypour SH, O'Brien C (2009) Optimizing whole supply chain benefit versus buyer's benefit through supplier selection. Int J Prod Econ 121:482–493
Kostamis D, Beil DR, Duenyas I (2009) Total-cost procurement auctions: impact of suppliers' cost adjustments on auction format choice. Manage Sci 55:1985–1999
Kumar A, Jain V, Kumar S (2014) A comprehensive environment friendly approach for supplier selection. OMEGA-Int J Manage Sci 42:109–123
Li C, Debo LG (2009) Second sourcing vs. sole sourcing with capacity investment and asymmetric information. Manuf Serv Oper Manage 11:448–470
Li X, Wang Q (2007) Coordination mechanisms of supply chain systems. Eur J Oper Res 179:1–16
Mansini R, Savelsbergh MWP, Tocchella B (2012) The supplier selection problem with quantity discounts and truckload shipping. OMEGA-Int J Manage Sci 40:445–455
Qi X (2007) Order splitting with multiple capacitated suppliers. Eur J Oper Res 178:421–432
Romeijn HE, Geunes J, Taaffe K (2007) On a nonseparable convex maximization problem with continuous knapsack constraints. Oper Res Lett 35:172–180
Tan T, Alp O (2009) An integrated approach to inventory and flexible capacity management under non-stationary stochastic demand and set-up costs. OR Spectr 31:337–360
Tan T, Alp O (2016) Optimal sourcing from alternative capacitated suppliers with general cost structures. OMEGA-Int J Manage Sci 58:26–32
Toptal A, Cetinkaya S (2008) Quantifying the value of buyer-vendor coordination: analytical and numerical results under different replenishment cost structures. Eur J Oper Res 187:785–805
Ustun O, Demirtas EA (2008) An integrated multi-objective decision-making process for multi-period lot-sizing with supplier selection. OMEGA-Int J Manage Sci 36:509–521
Xia Y, Chen B, Kouvelis P (2008) Market-based supply chain coordination by matching suppliers' cost structures with buyers' order profiles. Manage Sci 54:1861–1875
Zhang G, Ma L (2009) Optimal acquisition policy with quantity discounts and uncertain demands. Eur J Oper Res 47:2409–2425
Zhang J-L, Zhang M-Y (2011) Supplier selection and purchase problem with fixed cost and constrained order quantities under stochastic demand. Int J Prod Econ 129:1–7
1 Point Commerce (2014) http://www.1commerce.com

Chapter 9
Distribution, Transshipment, and Sustainable Logistics for Fashion Products

Zhi-Hua Hu

Abstract This paper reports and studies three fashion logistics optimization-related problems, including fashion distribution, transshipment, and sustainable logistics management. Fashion products are popular in the context of e-commerce and business innovation. The responsiveness of product searching, purchasing, experiencing, and returning or disposal is critically important. In this paper, considering the consumer experience during the distribution service, the try-on service is introduced and examined, where the effects of try-on services on return ratios, loyalty, and logistics cost are analyzed based on experiments. Then, to share fashion products among retailers of a brand, transshipment is critical and this paper presents an analytical model for investigating transshipment. By transshipment, fashion products can be efficiently transported from, e.g., the selling-well "good business" retailers to the "poor business" retailers. Next, considering the short life cycles of fashion products, a sustainable supply chain model is devised by using rental services in the loop. The operations problems and systematic mechanisms are then explored based on the model. Finally, the challenges and opportunities of using advanced technologies (e.g., Internet of Things and big data) are discussed. The proposed models in this paper provide important references for fashion industrialists as well as academics on fashion logistics and supply chain management.

Keywords Logistics management · e-commerce · Fashion logistics · Transshipment · Fashion supply chain · Green logistics · Fashion industry

Z.-H. Hu (✉)
Logistics Research Center, Shanghai Maritime University,
Shanghai 200135, China
e-mail: zhhu@shmtu.edu.cn

T.-M. Choi (ed.), *Analytical Modeling Research in Fashion Business*,
Springer Series in Fashion Business, DOI 10.1007/978-981-10-1014-9_9

9.1 Introduction

According to an industrial survey, more than 70 % consumers in China have experiences in buying fashion products (China Internet Network Information Center 2015) typically Vancl (www.vancl.com), Zara (www.zara.cn), Gap (www.gap.cn), and Uniqlo (www.uniqlo.cn). The fashion industry has expanded over the past few years. For example, the production capital of the industry in China in 2011 is about 3.2653 trillion yuan (RMB), increased by 6.34 times comparing to that of 2000, and increased 18.28 % a year on average. Before 2015, five famous international fashion brands (Uniqlo, H&M, Zara, Gap, and C&A) already established about nine hundred retail stores in China. For example, Uniqlo has operated 350 retail stores in 78 cities in China, especially in the second-tier cities.

Fashion is a "cross-sector concept" that encompasses several industries, such as apparel, footwear, leather, jewelry, perfumes, and cosmetics (Macchion et al. 2015). In this paper, the term "fashion" mainly refers to "fashion apparel." Many fashion firms now have tried to follow business strategies for reducing lead times, thereby responding rapidly to market trends in order to gain a competitive advantage (Barnes and Lea-Greenwood 2006b; Hines and Bruce 2001). By improving the efficiency of logistics, fashion products can be well distributed to retailers or final consumers, and transshipped among retailers. The increasing trend toward globalization and growing levels of international competition have enticed firms in the fashion industry to build strong and responsive supply chains that will enable efficient and effective operations and to offer customers the best value with a view to achieving market leadership (Ngai et al. 2014). Also, the fashion producers face sustainable development requests from environment organizations and society. In the fashion industry, many companies are adopting sustainable strategies on supply chain management and to balance among economic, environmental, and social performance. Logistics management provides an additional way to improve the sustainability of fashion products (Turker and Altuntas 2014).

A supply chain system of fashion distribution, transshipment, and sustainable logistics is presented in Fig. 9.1, which highlights the three core fashion logistics management–related problems to be investigated in the paper. The fashion products may be distributed from the distribution center to retailers or directly distributed to the final consumers. The fashion products can also be distributed to consumers from retailers. Among retailers, transshipment operations are used to share stocks by centrally controlled mechanisms. In this study, the sustainable logistics management, including the flows of used products from consumers to the material suppliers, is also examined.

The remaining parts of this paper are arranged as follows. First, in Sect. 9.2, a literature review on fashion distribution, transshipment, and sustainable logistics is presented. Then, Sect. 9.3 presents the analytical models. To be specific, following the three typical and important logistics operations as depicted in Fig. 9.1, we consider in the fashion distribution, transshipment, and sustainable logistics in

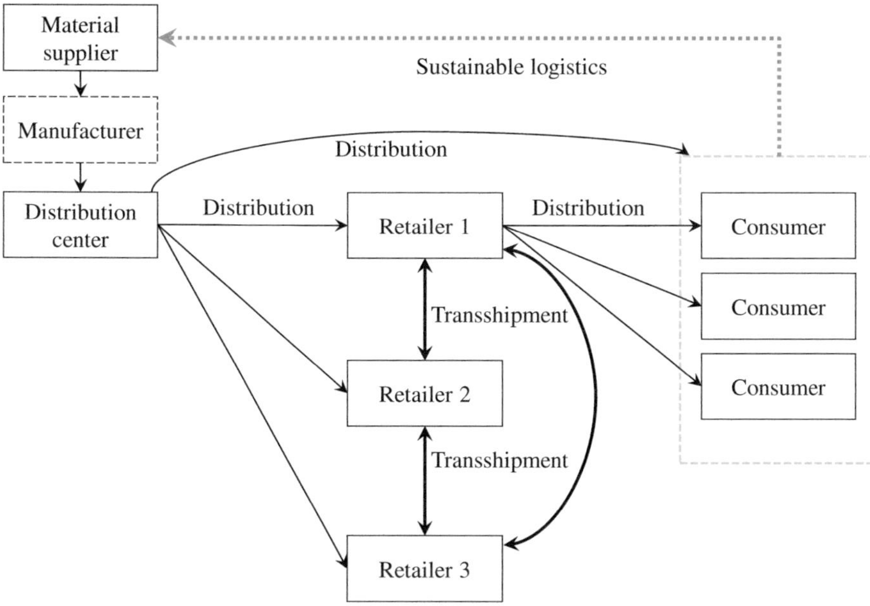

Fig. 9.1 A system of fashion distribution, transshipment, and sustainable logistics

Sects. 9.3.1, 9.3.2 and 9.3.3, respectively. Challenges and opportunities for using advanced technologies are discussed in Sect. 9.4. Section 9.5 concludes this paper.

9.2 Literature Review

9.2.1 Distribution

The vehicle routing problem (VRP) plays a central role in distribution management. Fashion distribution is related to the scope of city logistics, one key problem of which is the city VRP [see a delicate review by (Kim et al. 2015)]. Like the city VRP, during the traveling period of fashion distribution, traffic regulations, traffic congestion, road condition, parking space, air pollution, noise pollution, and emergencies should be considered. Also, four key stakeholders, shippers, carriers, residents, and administrators, are of interests. However, in the fashion distribution problem, consumers are special residents and shippers because the services to them affect the success of the distribution. Pillac et al. (2013) reviewed the dynamic VRPs in which the service times are dynamic. Eksioglu et al. (2009) presented a methodology for classifying the literature of the VRPs. They proposed that in terms of "on-site service/waiting times," four terms can be identified, namely "deterministic," "time dependent," "vehicle-type dependent," and "stochastic." In fashion distribution, the on-site service times are dependent on consumers and the service strategies (e.g., whether the try-on service is allowed or not). The distribution problem is subject to some types of VRP. So it is difficult to be solved, and many heuristics are invented. Prins (2004) presented a relatively simple but effective hybrid genetic algorithm. Potvin (2009) conducted a survey of the literature on applications of evolutionary algorithms (genetic algorithms, evolution strategies, and particle swarm optimization) for VRPs. In this study, the genetic algorithm approach is proposed to solve the fashion distribution problem.

9.2.2 Transshipment

Traditional inventory systems are organized on a hierarchical basis with transportation flows from one echelon to the next. Flexible systems also allow lateral transshipments, e.g., the horizontal collaboration, where inventory is shared between firms of the same level in the supply chain. In general, lateral transshipment research focuses on the practice of moving stock from retail outlets with inventory excess to retail outlets with inventory shortages, thus significantly saving holding costs at the former and shortage costs at the latter. Transshipments provide an effective mechanism for correcting discrepancies between the locations' observed demand and their available inventory and thus serve as a useful tool for utilizing inventory which has already been procured and delivered into the fashion supply chain system. The locations' replenishment can be coordinated and even combined with transshipment in order to avoid excessive procurement costs, such as fixed replenishment costs. Transshipments represent an effective pooling mechanism at the same echelon of a supply chain. As a result, transshipments may lead to cost reductions and improved services (through higher flexibility and responsiveness) in the supply chain without increasing inventories (Gong and Yücesan 2012).

Lateral transshipments support firms in dealing with demand variability and stock-outs. They can lead to a more balanced inventory system, too. In a review of the use of lateral transshipments in inventory management, 94 models were reviewed and classified (Paterson et al. 2011). Herer and Rashit (1999) addressed the problem of inventory management in a two-location inventory system, in which the transshipments are carried out as means of emergency or alternative supply after demand has been fully realized. Naseraldin and Herer (2011) provided a thorough analysis on a problem of optimally balancing strategic and operational decisions in the presence of lateral transshipments. The benefits of integrating location and inventory decisions with lateral transshipments were demonstrated in their study. Stanger et al. (2013) revealed how lateral transshipments support the blood supply chain to fulfill the requirements of being efficient and safe. Rosales et al. (2013) compared the transshipment among retailers after demand has been realized with the cost-free allocation to the retailers from the development of a centralized depot in terms of costs. Grahovac and Chakravarty (2001) found that sharing and transshipment of items often, but not always, reduce the overall costs of holding, shipping, and waiting for inventory. Unexpectedly, these cost reductions are sometimes achieved through increasing overall inventory levels in the supply chain. Gong and Yücesan (2012) used an infinitesimal perturbation analysis to examine the multilocation transshipment problem with positive replenishment lead times.

Transshipments within a supply chain can be difficult to implement as the costs and benefits are often incurred by different parties. This difficulty becomes even more problematic when the costs and benefits are not completely known by all parties (i.e., information asymmetry exists). Dong et al. (2012) introduced the role of asymmetric information into the design of supply chain transshipment contracts. They developed a multilevel contracting framework to encourage transshipments and improve performance in the presence of information asymmetry.

9.2.3 Sustainable Logistics

The fashion industry's environmental impact is very high, particularly in relation to its huge global volume: It accounts for 9.3 % of world's labors and 4 % of worldwide exports (Caniato et al. 2012). The fashion industry owes its special features and structure to its requirements for shortened lead times, faster inventory turnovers, and high-order fulfillment rates for customer demand at its peak points (Barnes and Lea-Greenwood 2006a). In the fashion industry, many companies are adopting sustainable strategies to redesign their supply chains to achieve proper coordination among stakeholders and to balance among economic, environmental, and social performance. Logistics provides an additional way to improve the sustainability of fashion products (Turker and Altuntas 2014).

In the realm of logistics and supply chain management, sustainability is studied in terms of reverse/green/low-carbon/closed-loop logistics/supply chain management generally. Green logistics, reverse logistics, and green supply chain management have been the subject of numerous survey papers in both qualitative and quantitative disciplines (see Dekker et al. 2012; Agrawal et al. 2015; Fahimnia et al. 2015). Sustainable logistics and supply chains have attracted attention from researchers. Eskandarpour et al. (2015) analyzed 87 papers in the field of supply chain network design, covering mathematical models that include pertinent economic factors as well as environmental and/or social dimensions, while Touboulic and Walker (2015) conducted a structured literature review to map the use of theories in the field.

Sustainability is crucial nowadays and the design, planning and operations of sustainable logistics systems are challenging for the involved companies (Ramos et al. 2014). Xifeng et al. (2013) developed a multiobjective optimization model for sustainable facility location selection problem with the trade-off among economic, service, and environmental factors, especially the carbon dioxide emissions of transport. Digiesi et al. (2012) proposed an inventory model to support decision making in means of transport selection and order lot sizing (sustainable order quantity) which minimizes both logistics and environmental costs. Here, the dependency of logistical and environmental costs is shaped as functions of the 'loss factor,' a parameter adopted to classify means of transport based on loss in transport energy. García-Arca et al. (2014) implemented the approach for sustainable packaging logistics to improve the competitiveness of the supply chain. Bing et al. (2014) optimized the design of a reverse logistics system, which makes the overall recycling system more efficient and sustainable under the minimization of transportation cost and environmental impact. Bouhouras and Basbas (2015) presented policies and measures based on successful case studies that were adopted in order to control the environmental impacts of the city logistics operations.

9.3 The Three Typical Operations: Models and Analysis

9.3.1 Distribution

E-commerce is a popular choice as a supplement to the off-line channel in the fashion industry. Many famous fashion brands have owned both online and off-line channels. In particular, many apparel customization services are provided to young people. In China, many online apparel customization merchants are active after e-commerce has become popular. Logistics is a way to ensure customer satisfaction by experiencing the services from the merchant. The try-on service when apparel is distributed becomes a focus of such services (e.g., www.vancl.com). Therefore, specialized deliverers are required and they provide try-on services directly to customers when they deliver the apparel to them. However, dilemma exists in such a service. To be specific, the try-on service will cost time, e.g., 10–20 min, such that the logistics cost increases. Moreover, the times by the try-on service are usually uncertain, such that the following customers in the same route may be delayed. However, two direct merits are identified. First, after having the try-on service, the possible returns can be directly collected by the deliverer, such that the related logistics cost is saved. Second, by try-on service, the customer satisfaction degree is improved such that the customer will keep buying apparel from the merchants ever though she returns the apparel this time. Therefore, the customer loyalty is improved. This work studies the impacts of apparel distribution with try-on service on logistics cost and customer satisfaction.

This part relates and contributes to two streams of research in the literature: fashion apparel e-commerce operations and routing problems. The apparel e-commerce literature focuses on related Web technologies, consumer buying behaviors, and supply chain management. Few studies aim at logistics services although proper logistics management guarantees the efficacy of apparel e-commerce. Routing optimization is a very active research area, and many studies are published in the literature. Uncertain demands and traveling times are researched. However, stochastic service times at customers are scarcely considered. This study identifies the problem of try-on service in apparel e-commerce and logistics. A real-world distribution network is used, and an evolutionary algorithm is devised to reveal the features of the problem. The impacts of uncertain try-on service times on the logistics cost and customer satisfaction are examined by computational experiments. The implications of the methods to the apparel distribution practitioners are discussed.

9.3.1.1 Formulation

The apparel distribution model is built upon traditional capacitated VRP (CVRP) models. The set $V = \{1, 2, \ldots, n, n+1, n+2\}$ presents the customers and indexed by i and j, where 1 and n + 2 are two overlapping depots, denoted by D^1 and D^2. Therefore, the customer set is denoted by $N = V \backslash \{1, n+2\}$. The vehicle set (or deliver set) is denoted by O and indexed by o. The try-on time at customer i is denoted by P_i. The traveling time from customer i to j is denoted by $T_{i,j}$. The estimated try-on time at customer i is denoted by S_i. The maximum work time is denoted by U. A "big number" is denoted by M. The problem is to obtain the optimal routes. Therefore, three groups of decision variables are defined. First, $x_{oij} \in \{0, 1\}$ is a flow variable. When a deliver o serves customer j just after serving customer i, $x_{oij} = 1$; otherwise, zero. Third, $t_{k,i}$ is the time that the deliver o distributes apparel products to customer i. Because apparel products are usually light and their occupied storages are limited, the volume constraints in the traditional CVRP are not considered in this study.

The model with deterministic try-on times is formulated in Eqs. (9.1)–(9.7). The objective Eq. (9.1) minimizes the total traveling time. By Eqs. (9.2)–(9.7), six groups of constraints are defined. First, every customer can be served only one time (Constraint (9.2)). Second, the flow at each customer should be balanced (Constraints (9.3)–(9.5)). Every deliver starts from the depot and finally returns to it (Constraints (9.3)–(9.4)). The flows are balanced for every customer (Constraint (9.5)). Third, the visit time at each customer is constrained by Constraint (9.6). Every deliver returns to the depot during the work time.

$$\min f = \sum_{o,i,j} \left(T_{ij} \cdot x_{oij} \right) \tag{9.1}$$

$$\sum_{oj} x_{oij} = 1, \quad \forall i \in N \tag{9.2}$$

$$\sum_{oj} x_{oD^1j} = 1, \quad \forall o \tag{9.3}$$

$$\sum_{i} x_{oiD^2} = 1, \quad \forall o \tag{9.4}$$

$$\sum_i x_{oij} = \sum_i x_{oji}, \quad \forall o, \quad j \in N \tag{9.5}$$

$$t_{oi} + S_i + T_{ij} - t_{oj} \leq \left(1 - x_{oij}\right) \cdot M, \quad \forall o, i, j \tag{9.6}$$

$$t_{oD^2} \leq U, \quad \forall o \tag{9.7}$$

The above proposed model is a mixed-integer linear programming model, which is an extension of VRP and CVRP models. Therefore, it can be solved by column generation, branch and bound, Lagrangian relaxation, or other hybrid exact algorithms. It can also be solved by designing the evolutionary algorithm, simulated annealing algorithm, Tabu search, or other heuristic algorithms. In the following, an evolutionary algorithm is designed for medium- and large-scale instances.

Try-on time is a stochastic parameter with a statistical distribution that can be determined by analyzing historical data. Here, the uniform distribution is assumed, denoted by $U[a, b]$ (unit: minute). Three strategies are devised to cope with the uncertain try-on times when a routing solution is planed and when the try-on times are to be realized at customers. First, the try-on time is estimated with different risk preferences. An estimation coefficient (α) refers to this preference (e.g., 0.5 for moderate, 0.75 for reserve, and 0.25 for optimistic). Based on this parameter, the try-on time is estimated by $S^E = a + \alpha \cdot (b - a)$. Second, a period of time can be reserved to cope with the try-on time uncertainty. By the reservation ratio of duration for uncertainty (β), the duration of route is denoted by $U^T = (1 - \beta) \cdot U$. Third, when the uncertain try-on time is to be realized at a customer, and the rest of time is not sufficient for the remainder customers, the strategy of try-on rejection is to be made.

In order to reveal customer satisfaction and induce customer loyalty, the following scenarios are considering by introducing new parameters. When a customer tries an apparel, if she is not satisfied by the received apparel, she will return it to the deliverer and then no return cost is incurred, and her sale will be lost with the probability P^{LaT} (lost after try-on, LAT); if she does not experience the try-on service, her sale will be lost with the probability P^{LnT} (lost without try-on, LNT). Apparently, $P^{\mathrm{LnT}} > > P^{\mathrm{LaT}}$. Because the buying decisions made by customers are online purchase decisions, the return ratio is set to P^{RoT} (ratio of return). The return ratio is not affected by the try-on service experiences. The ratio of return cost to unit distance transportation cost is denoted by σ, which is determined by investigation. Therefore, a parameter vector with the above parameters is denoted by $P = (\alpha, \beta, \sigma, P^{\mathrm{RoT}}, P^{\mathrm{LaT}}, P^{\mathrm{LnT}})$.

Considering the try-on time uncertainty, four assessment criteria of routing solution and their notations are proposed. The distribution cost presented by transportation cost of all routes is denoted by f^D; the ratio of customers not experienced the try-on service is denoted by f^O; the return cost for unsatisfied apparel is denoted by f^R; finally, the ratio of lost customers indicated by the solution is denoted by f^S. Here, the distribution and return costs are aggregated into logistics cost, e.g., $f^C = f^D + f^R$. Conceptually, the service failure is presented by ratio of lost customers (f^S). Therefore, an assessment criteria vector is denoted by $F = (f^D, f^O, f^R, f^S, f^C)$.

9.3.1.2 Solution Methodology

An evolutionary algorithm (Algorithm 9.1) is devised to find the optimal solution for VRP with deterministic try-on service time and maximum traveling time limits.

Algorithm 9.1 (The evolutionary algorithm)

Input: The population size, P_s; crossover probability, P_x; mutation probability, P_m; maximal iterations, P_g; the sets; and parameters defined in the model in Sect. 9.3.
Output: Optimal routes
Process

Step 1 Initialization: generate a population P with P_s individuals that are generated randomly; the elite is set, $elite = \Phi$.
Step 2 Evaluation: compute the fitness of each individual by Eq. (9.1).
Step 3 Selection: choose $\lceil P_s/2 \rceil$ pairs of individuals by the tournament selection strategy.
Step 4 Crossover: for each pair of the $\lceil P_s/2 \rceil$ pairs, by the probability P_x, apply the partially mapped crossover (PMX) operator to produce new individuals and replace the parents.
Step 5 Mutation: by the probability P_m, perform mutation on new individuals by swapping two genes randomly.
Step 6 Evaluate the new population by Step 2.
Step 7 Update *elite* by the best individual.
Step 8 If the termination conditions are not satisfied, go to Step 3; or else return *elite*.

1. Encoding and decoding. The sequence of the customers (N) is used for encoding, where each customer represents a gene. Therefore, any permutation of the sequence is a valid chromosome. The decoding procedure should construct the route set. In this study, the decoding strategy adopted is based on the shortest path algorithm for multitype goods distribution, which is studied by Liu et al. (2008). In the following, a revised decoding approach is proposed. The route represented by a customer sequence $r = \langle c_1, c_2, \ldots, c_k \rangle$ is calculated by Eq. (9.8). A permutation of the customer set N is denoted by γ, which is updated by inserting the depot indexes as the first and last components. In order to disconnect the two overlapped depot indices, set $T_{1,n+2} = \infty$. The subsequence from the $(i+1)$ component to the j component is denoted by $\gamma_{i,j}$. Moreover, $C_{i,j}$

simply represents $C(\gamma_{i,j})$ and is computed by Eq. (9.9). Therefore, for a given route γ representing a sequence of customers, $C_{i,j}$ representing the "distance" among any two "nodes," the set of roads from the first "node" (starting from the depot) to the last "node" (returning to the depot) represents the set of routes with the minimum travel and service time.

$$C(r) = T_{D^1c_1} + \sum_{i\in\{1,2,\ldots,k-1\}} \left(T_{c_ic_{i+1}} + S_{c_i}\right) + T_{c_kD^2} \tag{9.8}$$

$$C_{i,j} = \begin{cases} C(\gamma_{i,j}), & C(\gamma_{i,j}) \le U \\ \infty, & \text{else} \end{cases} \tag{9.9}$$

2. The selection and crossover operators: The tournament selection strategy (Ryvkin 2010) is adopted to generate new pairs, followed by a partially matched crossover (PMX) (Ting et al. 2010) operation. The probability of crossover is controlled by P_x. The new pairs then update the original parents by replacing them.
3. The mutation operator: The population updated by selection and crossover operations endures a mutation process by swapping any two randomly chosen genes by the mutation probability (P_m).
4. The control parameters: The evolutionary process is controlled by the following parameters: population size (P_s), crossover probability (P_x), mutation probability (P_m), and maximal iterations (P_g). Prins (2004) studied the parameter configuration of an evolutionary algorithm for VRP. He found that the encoding and decoding strategies, the evolutionary operators, and the assessment criteria for cost and other measures all would impose important effects on the convergence and evolution speed.

The above algorithm is extended to deal with the try-on time uncertainty. The try-on service time is estimated, the estimation value ($S^E = a + \alpha \cdot (b - a)$) replaces S_i in Eq. (9.8), and the tour traveling time limit (U) is replaced by $U^T = \beta \cdot U$. Furthermore, in the fitness assessment part, by simulating the try-on time for a specific time (denoted as P^{sim}), the expected values of $F((F = (f^D, f^O, f^R, f^S, f^C)))$ are computed and recorded. When a route is simulated, the strategies in Sect. 9.3.2 apply.

9.3.1.3 Expriments and Results

Totally 582 nodes including a depot node are considered in the dataset that is a simplified road network of Lujiazui area at Pudong District of Shanghai, China. Figure 9.2 shows the nodes and roads. The customer nodes are selected from the network. The distance matrix is computed by a geographical information system.

The purpose of the experiments is to reveal the effects of solution strategies and parameters, with $P = (\alpha, \beta, \sigma, P^{\text{RoT}}, P^{\text{LaT}}, P^{\text{LnT}})$ on the five criteria of routes,

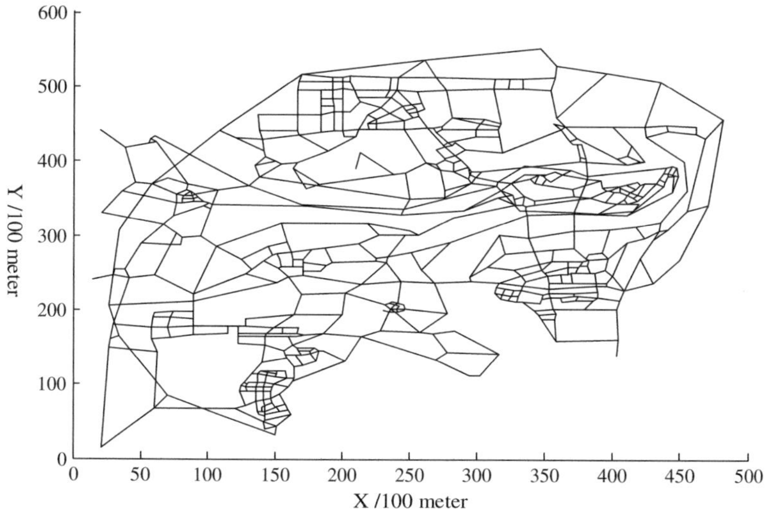

Fig. 9.2 A road network for test

$F = (f^D, f^O, f^R, f^S, f^C)$. Five scenarios are designed for this purpose. In the experimental study, a dataset with 100 customers is used; the general working time of a day is eight hours, $U = 8$; the simulation time for a solution of prior routes is set to $P^{\text{sim}} = 500$. The default values of settings are set to $P = (0.5, 1, 0.4, 0.5, 0.1, 0.7)$. The algorithm is implemented by MATLAB 8.1 and run on a personal computer with Intel ® Core (TM) 2 Duo 2.26 GHz CPU and 2.93 GB RAM. The parameters of the evolutionary algorithm are set to be the following: population size, $P_s = 20$; crossover probability, $P_x = 0.7$; mutation probability, $P_m = 0.1$; and maximal iterations, $P_g = 200$. The experiment steps are shown in Table 9.1.

The following discussion summarizes the numerical results and the procedures performed in the five scenarios based on the results in Figs. 9.3, 9.4, 9.5, and 9.6, and Tables 9.2 and 9.3. The results of experimental scenarios in the figures and tables are discussed below.

Experimental scenario 1: Fig. 9.3 depicts the routing results with thick lines, and Table 9.2 shows $(f^D, f^O, f^R, f^S, f^C)$ for each route and a summary of them. By the default settings of $(\alpha, \beta, \sigma, P^{\text{RoT}}, P^{\text{LaT}}, P^{\text{LnT}})$, comparing to transportation cost (f^D), the return cost (f^R) is low $(<1\%)$ and zero for three routes; most customers experienced try-on service such that f^O is also low and zero for three routes; furthermore, few customers will be lost indicated by f^S.

Experimental scenario 2: Fig. 9.4 presents the curves of F $(f^D, f^O, f^R, f^S, f^C)$ when the estimation coefficient (α) varies from 0 to 1 by a step of 0.05. When α increases, according to the formula, $S^E = a + \alpha \cdot (b - a)$, the estimated try-on time increases. Because the working time of a day (U) is fixed, the time available for traveling is

Table 9.1 Experiment scenarios

No.	Purpose and description
1	Demonstrate the routes and services 1. Draw the routes 2. Calculate F for all routes in a table
2	Study the effect of α 1. Set $\alpha = 0, 0.05, 0.1, \ldots, 1$; 2. Record F for all routes when α varies; 3. Draw curves for F when α varies
3	Study the effect of β 1. Set $\beta = 0, 0.05, 0.1, \ldots, 1$; 2. Record F for all routes when β varies; 3. Draw curves for F when β varies
4	Conduct the sensitivity analysis for parameters $\sigma, P^{\mathrm{RoT}}, P^{\mathrm{LaT}}, P^{\mathrm{LnT}}$. Study the impacts of that a parameter varies for −75, −50, −25, +25, +50, +75 % on the varying percentages of five objectives
5	Conduct the Pareto analysis when f^S, f^C are objectives Vary α, β and record all pairs of f^S and f^C and draw the Pareto front

Note $F = (f^D, f^O, f^R, f^S, f^C)$; $P = (\alpha, \beta, \sigma, P^{\mathrm{RoT}}, P^{\mathrm{LaT}}, P^{\mathrm{LnT}})$

Fig. 9.3 Demonstration of a solution with four routes

reduced. This reduction directly explains the increasing transportation cost. Within increasing α, sufficient time is reserved for try-on and return, such that the return cost diminishes. Comparing to the transportation cost, the return cost is minor. Therefore, the total cost is also increasing with α. Similarly, the ratio of customers not experiencing the try-on service (f^O) and the ratio of lost customers indicated by the solution (f^S) decrease with the increase of α. Specially, when α reaches 0.5, the

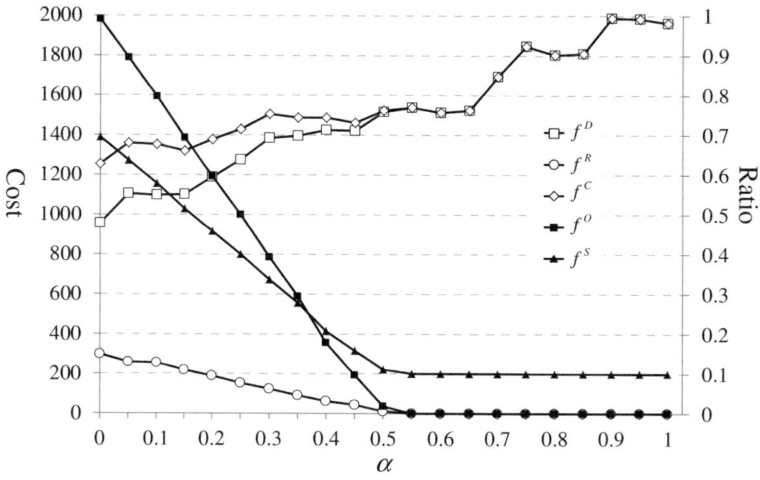

Fig. 9.4 Impacts of try-on time estimation coefficient on F (No. 2)

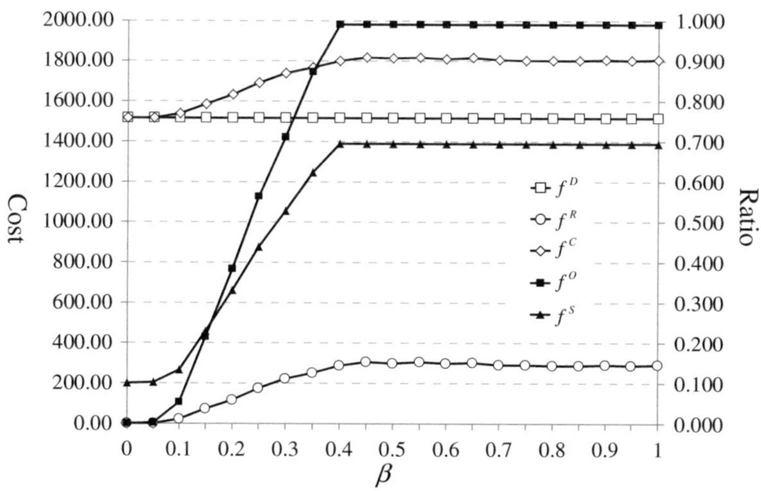

Fig. 9.5 Impacts of time reservation ratio on F (No. 1)

two ratios (f^O and f^S) get stable, where the lost customer ratio is 0.1 and almost all customers get the try-on service.

Experimental scenario 3: Fig. 9.5 presents five curves when the reservation ratio of duration for uncertainty (β) increases. The time reserved will reduce the time available for transportation such that the transportation cost and the total cost increase. The increase of reserved time for the uncertain try-on time also helps reduce the decreases of return cost.

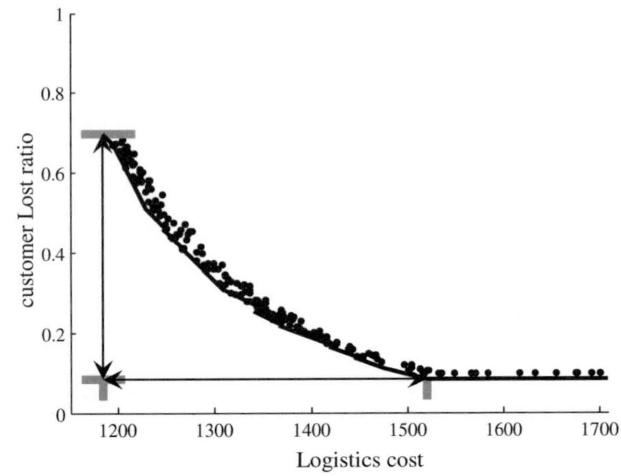

Fig. 9.6 The Pareto frontier between the customer lost ratio and the logistics cost (No. 5)

Table 9.2 Assessing the routes and solution (No. 1)

Route	Travel time/mi	f^D/yuan	f^O (%)	f^S (%)	f^R/yuan	f^C/yuan
1	349.16	311.38	5	13	3.58	314.96
2	351.53	313.49	4	13	4.84	318.33
3	326.89	293.81	0	10	0	293.81
4	329.73	292.05	0	10	0	292.05
5	339.27	302.56	0	10	0	302.56
Sum	1696.58	1513.01	1.8	11	8.42	1521.43

Experimental scenario 4: Table 9.3 presents the results of sensitivity analysis for $\sigma, P^{\text{RoT}}, P^{\text{LaT}}, P^{\text{LnT}}$. Because $P^{\text{RoT}}, P^{\text{LaT}}, P^{\text{LnT}}$ are values between 0 and 1, "−a %" means that the parameter is adjusted to be $a\,\% \cdot P^*$, whereas "a %" means $P^* + a\,\% \cdot (1 - P^*)$. This rule ensures that P^* drops between 0 and 1. Table 9.3 reveals three rules. First, the return costs are sensitive to all parameters more than other four criteria. And therefore, the total costs are a little sensitive to all parameters because comparing to transportation costs the return costs are minor. Second, the ratio of lost customers (f^S) is sensitive to P^{LaT} and P^{LnT} because these two parameters set the influence degrees of lost customers (when try-on services are used or not). Third, the impacts on f^D and f^O are not clear in the experiments.

Experimental scenario 5: Fig. 9.6 presents the trade-off between the logistics cost and the lost customer ratio. Apparently, they are conflicting objectives. In order to decrease the lost customer ratio from 0.7 to 0.07, an additional half cost should be paid (about from 1180 to 1520). This tool provides a quantification method to simultaneously consider the cost and the service quality.

Table 9.3 Results of sensitivity analysis for four parameters

				σ							p^{RoT}			
	−75 %	−50 %	−25 %	0	+25 %	+50 %	+75 %	−75 %	−50 %	−25 %	0	+25 %	+50 %	+75 %
f^O	0	0	0	0.25	−4	0	0	−4	0	0	0.25	0	0	0
f^D	0	0	0	1535	0	0	0	0	0	0	1535	0	0	0
f^R	**−75**	**−49**	**−24**	201	**16.6**	**50.9**	**67.2**	**−76**	**−52**	**−21**	194	**25.6**	**56.6**	**80.9**
f^S	0	0	0	0.25	0	0	0	0	0	0	0.25	0	0	0
f^C	−8.7	−5.7	−2.8	1736	1.92	5.9	7.79	−8.6	−5.8	−2.4	1729	2.87	6.35	9.07
				p^{LaT}							p^{LnT}			
	−75 %	−50 %	−25 %	0	+25 %	+50 %	+75 %	−75 %	−50 %	−25 %	0	+25 %	+50 %	+75 %
f^O	−4	0	0	0.25	0	0	0	−4	0	0	0.25	0	0	0
f^D	0	0	0	1535	0	0	0	0	0	0	1535	0	0	0
f^R	**−0.7**	**7.81**	**9.18**	193	**7.41**	**7.29**	**-1.6**	**−1.8**	**−9.9**	**−6**	207	**−5.6**	**−4.1**	**−10**
f^S	**−20**	**−16**	**−8**	0.25	**8**	**16**	**24**	**−52**	**−36**	**−16**	0.25	**20**	**32**	**32**
f^C	−0.1	0.87	1.02	1728	0.83	0.81	−0.2	−0.2	−1.2	−0.7	1742	−0.7	−0.5	−1.2

9.3.1.4 Discussions on Distribution

This part identifies a new distribution and service problem arising from fashion industry. The problem has distinct features that are not studied in the literature. The important try-on service is considered during the apparel distribution process. The try-on time uncertainty is coped with by three strategies: adjusting the try-on time estimation parameter, adjusting the time reservation ratio for each route, and applying the try-on service rejection when distribution time is insufficient for remainder customers in the route. An evolutionary algorithm is devised and implemented for optimizing routes with the minimum logistics cost and simulating uncertain try-on times. By simulation, the try-on service ratio, the return cost, and the probability of losing customers are assessed. Five experimental scenarios are performed to study the impacts of parameters and solution strategies on the cost and service criteria. Based on the formulations, algorithm, and experimental results, a decision support system (DSS) is designed to integrate the apparel e-commerce Web portal, customer and order management, mapping tool, and interactive decision process. The DSS provides tools to analyze and design distribution strategies and tuning parameters for try-on services and the customer satisfaction affected by the try-on services. However, the parameter estimation and the interactive decision methods should be further studied because they determine the preferences of optimization criteria and the control of try-on servicing strategies with uncertain try-on times.

9.3.2 Transshipment

The fashion industry is one of the most important industries in the world, and it has rapidly developed over the past decade (Choi et al. 2013). Under the fast fashion trend, fashion companies aim at reducing lead times to get products promptly from conceptual design to retail sales floor (Barnes and Lea-Greenwood 2006a; Sull and Turconi 2008). Notice that the importance and principles of the fast fashion industry have been well studied in the literature of supply chain management, and consumer psychology and behaviors. Even though demand management such as fashion sales prediction, fast design and manufacturing technologies, and quick response practices are widely studied in the literature in the context of fast fashion, the needed logistics supports (such as transshipment) that connect all the industrial processes and parties together seem to be neglected (although some researchers focusing on supply chain management usually suggested that logistics operations affect the degree of quick response [Barnes and Lea-Greenwood 2010; Cheng et al. 2013; Choi et al. 2011; Eliiyi et al. 2011; Tokatli et al. 2008)].

In this subsection, we consider the fashion transshipment problem (FATP) among retailers in an alliance for fashion sales. Transshipment is a practical way to share stocks among retailers. However, the spontaneous transshipment between only a few (generally two) retailers fails to utilize the in-transit stocks and thus fails

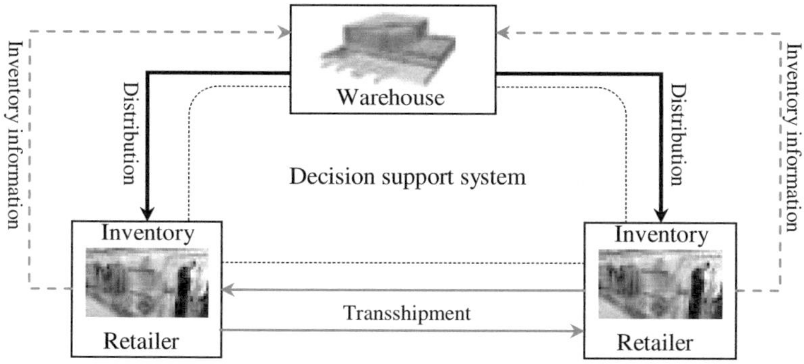

Fig. 9.7 The transshipment problem of apparel products among retailers

to improve the profit because of information sharing problems. Moreover, in practice, this kind of transshipment strategy is forbidden by the upstream dealers or the brand owners because the downstream transshipments will be out of control and the sales prices may be pushed up by some retailers that hold the stocks. Here, alliance refers to the controllable sharing of stocks among retailers. Transshipment may not be connected with the lead time of fashion apparel products, whereas it helps to reduce stocks and the dependences on sales prediction. Apparently, effective transshipment depends on information sharing mechanisms among retailers, and it should be implemented as a decision support system (DSS) that optimizes the transshipment solutions.

In the research community of transportation, transshipment is the shipment of goods or containers to an intermediate destination and then to another destination. In this work, transshipment is to transport fashion apparel products from one retailer to another retailer, rather than from a distribution center to the retailers. Commonly, the excessive stocks are collected by a distribution center and then redistributed to other retailers or become stocks at the distribution center. Figure 9.7 presents the flows of inventory information and apparel products among the distribution centers and the retailers. A decision support system is used to coordinate and optimize the information and product flows.

9.3.2.1 Formulation

In the studied fashion apparel transshipment problem (FATP), the set of retailers is denoted by $R = \{1, 2, \ldots, N^R\}$ and indexed by r and e. To facilitate the formulation, a finite set of number of apparel products introduced is denoted by $Q = \{0, 1, \ldots, N^Q\}$ and indexed by q. The logistics network among retailers is the base of the transshipment optimization. The distance from the retailer r to e in the network is denoted by $D_{r,e}$. Based on the sales historical data and the present stock held by a retailer r (denoted by H_r), the retailer can predict or estimate its sales for the next time period. The estimations are represented by statistical distributions. Directly, the probability of the retailer r that will sell q apparel products is taken as input data, denoted by $P_{r,q}$. The values of $P_{r,q}$ are estimated by marketing information and other sources of information for the next period of sales. However, this work only builds a model for the present period. Four unit costs and profit parameters are defined as follows. First, C^- denotes the unit penalty cost of shortage of stock comparing to the demand; C^+ denotes the unit penalty cost of excess of stock comparing to the demand; C^T denotes the unit cost of transshipment for transporting an item of apparel for a kilometer distance; and C^P is the profit of selling an apparel product. In the transshipment process of apparel products, packages, binding, and special requirements for apparel products of high-quality demand, additional handling costs are considered in a unified way by the above parameters.

The FATP is to determine the quantity of apparel products that are transshipped from retailer r to retailer e, which is denoted by $x_{r,e}$. h_r denotes the quantity of apparel products held by retailer r after transshipment. $h_{r,q}$ denotes the quantity of apparel products held by the retailer r after transshipment when the retailer estimates q apparel products for sales in the next period. In other words, q represents the uncertain scenario. Therefore, $h_{r,q}$ depends on the difference between q and h_r. Based on $h_{r,q}$, $h^+_{r,q}$ and $h^-_{r,q}$ denote the excess and shortage of apparel products corresponding to the uncertain scenario indicated by that q apparel products are to be sold in the next period by the retailer r. Considering the exact numbers of apparel products, the above variables are all integers.

In the transshipment decision, four criteria are involved. The expected profit considering the uncertain numbers of apparel products to be sold should be maximized, denoted by f^P; the transportation cost of transshipment, f^T, should be minimized; the penalty cost for the stocks exceeding the estimated demands, f^+, should be minimized; similarly, the penalty cost for shortage comparing to the estimated demands, f^-, is to be minimized. Because the above four criteria all can be defined by monetary costs, the formulated model maximizes the profit subtracting the costs. Based on the above notations and analysis, [M1] is formulated by Eqs. (9.17)–(9.19).

$$[\mathrm{M1}]\max f = f^P - f^+ - f^- - f^T$$

$$\text{where}\quad f^P = \sum_{r,q} P_{r,q}(q, h_r)^- C^P = \sum_{r,q} P_{r,q} h_{q,r} C^P \tag{9.10}$$

$$f^+ = \sum_{r,q} P_{r,q}(h_r - q)^+ C^+ = \sum_{r,q} P_{r,q} h^+_{r,q} C^+ \tag{9.11}$$

$$f^- = \sum_{r,q} P_{r,q}(q - h_r)^+ C^- = \sum_{r,q} P_{r,q} h^-_{r,q} C^- \tag{9.12}$$

$$f^T = \sum_{r,e} D_{r,e} C^T x_{r,e} \tag{9.13}$$

s.t.

$$h_r = H_r + \sum_e x_{e,r} - \sum_e x_{r,e} \quad \forall r, \tag{9.14}$$

$$h^+_{r,q} \geq h_r - q, \forall r, q \tag{9.15}$$

$$h^-_{r,q} \geq q - h_r, \forall r, q \tag{9.16}$$

$$h_{r,q} \leq q, \quad \forall r, q \tag{9.17}$$

$$h_{r,q} \leq h_r, \quad \forall r, q \tag{9.18}$$

$$x_{r,e}, h_r, h_{r,q}, h^+_{r,q}, h^-_{r,q} \geq 0, \quad \forall r, e, q \tag{9.19}$$

In the derivation formulas of f^P, f^+ and f^- in [M1], $(a, b)^- = \min(a, b)$, and $(a, b)^+ = \max(a, b)$; further, $(a)^- = (a, 0)^-$, and $(a)^+ = (a, 0)^+$. Therefore, by introducing $h_{r,q}$, $h^+_{r,q}$ and $h^-_{r,q}$, the nonlinear expressions of $(a, b)^-$ and $(a, b)^+$ are linearized, as shown in Eqs. (9.15)–(9.18). Equation (9.13) computes the transportation cost of the transshipment solution. In terms of stochastic programming, [M1] is a two-stage optimization model, where minimization of f^T is the first stage, whereas the maximization of $f^P - f^+ - f^-$ is the second stage under the stochastic scenarios. Equation (9.14) balances the inbound and outbound flows of apparel products at each retailer. Finally, the constraints of variable values are denoted by Eq. (9.19).

To assess the solution without transshipment under uncertain sales scenarios, [M2] is developed by simply setting $x = 0$ for all pairs of retailers. Apparently, $f^T = 0$ in [M2].

$$[\text{M2}] \max f = f^P - f^+ - f^- - f^T$$

s.t. Equations (27)–(36)

$$x_{r,e} = 0, \quad \forall r, e \tag{9.20}$$

9.3.2.2 Experiments and Results

This work uses the case of a famous Chinese fashion apparel brand. Figure 9.8 presents the distribution of its 43 retailers in Shanghai, China. Table 9.4 presents the addresses of the retailers. The distance matrix is estimated by the mapping

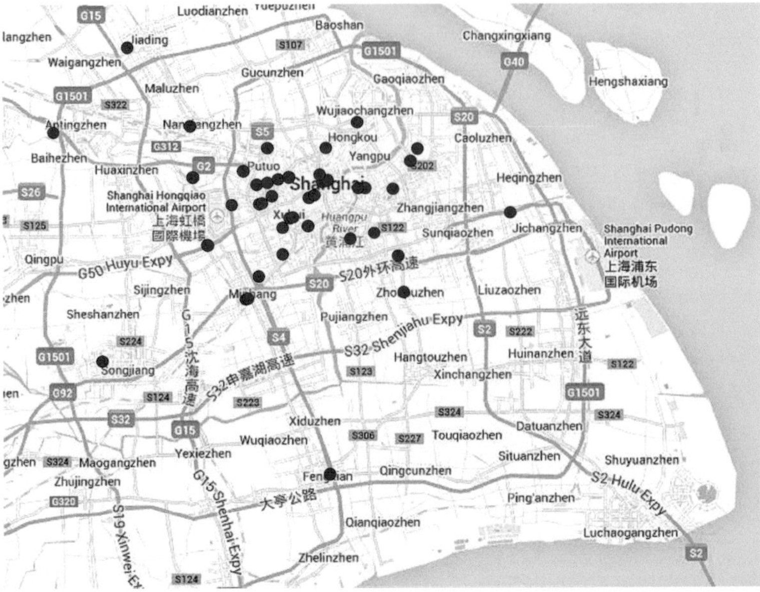

Fig. 9.8 Distribution of the retailers (an example)

services provided by *map.baidu.com*. Table 9.5 presents a part of the matrix. This case considers the simple uniform distribution of estimated sales in the next period for each retailer. The present stocks and the boundaries of the uniform distributions are provided in Table 9.6. In Table 9.6, H_r is the present stock level, and $\underline{V}$ and $\overline{V}$ are the estimated lower and upper limits of demands, which are used for the parameters of uniform distributions.

Three groups of experiments were performed to demonstrate the transshipment solutions and employed to analyze the effects of transshipment on solutions and the parameter sensitivities.

1. **Demonstrate the transshipment solutions**

By using the data estimated above, [M1] was solved with the results, $f = 513140.2$, $f^P = 522432.07$, $f^T = 2614.7$, $f^+ = 3763.59$, and $f^- = 2913.59$. Table 9.7 presents the 39 pairs of retailers and the transshipped amounts of apparel products. For example, from the retailer #1 (Dongfang Road #796), seven and two apparel products are transshipped to the retailers #5 and #6 (Siping Road #2500, and Zhangyang north Road #801), respectively. Totally 19 retailers transship some apparel products to other retailers, and 319 apparel products are transshipped. Among all retailers, the retailer #37 transships 41 apparel products to other retailers, and the retailer #27 accepts 43 apparel products from other retailers. These two retailers have the biggest "output" and "input" among all retailers.

Figure 9.9 shows the present stock, stock after transshipment, and the estimated boundaries of sales of apparel products for each retailer for the next period. In this

Table 9.4 Retailers and their addresses (an example)

No.	Address
1	Dongfang Road #796
2	Zhangyang Road #501
3	Hunan Road #2420
4	Zhangyang Road #3611
5	Siping Road #2500
6	Zhangyang north Road #801
7	Huangdian Road #300
8	Nianjiabang Road 518
9	Chengshan Road #1993
10	Chengshan Road #500
11	Huaxia East Road #2255
12	Baiqi Road #288
13	Middel Huahai Road #918
14	Middel Huahai Road #658-666
15	Hongqiao Road #1
16	Zhaojiabang Road #1000
17	Yishan Road #455
18	South 2nd Zhongshan Road #699
19	Huming Road #7388
20	Doushi Road #5001
21	Doushi Road #5001
22	Huming Road #9001-3
23	Qixin Road #3755
24	Zhongshan North Road #3300
25	Jinshajiang Road #788
26	Zhenguang Road #1288
27	Wuning Road #101
28	Changshou Road #401
29	Zhenhua Road #888
30	Tianshan Road #900
31	Xianxia West Road #88
32	Tianshan Road #789-889
33	Changning Road #1018
34	Nanjing East Road #588
35	Nanjing East Road #409-459
36	Nanjing East Road #819
37	Xizang North Road #166
38	Xijiangwan Road #388
39	Chengzhong Road #66
40	Moyu South Road #1077
41	Zhengnan Road #4368
42	Jingshajiang West Road #1051
43	Xinsongjiang Road #925

Table 9.5 Part of the distance matrix between retailers (km)

$D_{r,e}$	1	2	3	4	5	6	7	8	9	10	…
1	0	0.75	8.96	6.07	7.96	7.6	3.09	13.27	5.48	6.42	…
2	0.75	0	9.32	6.69	7.87	8.17	3.84	13.57	5.7	6.31	…
3	8.96	9.32	0	11.53	16.71	13.15	8.08	4.44	3.87	5.72	…
4	6.07	6.69	11.53	0	7.59	1.71	3.94	15.86	9.6	11.66	…
5	7.96	7.87	16.71	7.59	0	7.47	8.95	21.12	13.43	14.12	…
6	7.6	8.17	13.15	1.71	7.47	0	5.65	17.43	11.31	13.36	…
7	3.09	3.84	8.08	3.94	8.95	5.65	0	12.51	5.69	7.75	…
8	13.27	13.57	4.44	15.86	21.12	17.43	12.51	0	7.89	8.74	…
9	5.48	5.7	3.87	9.6	13.43	11.31	5.69	7.89	0	2.8	…
10	6.42	6.31	5.72	11.66	14.12	13.36	7.75	8.74	2.8	0	…
…	…	…	…	…	…	…	…	…	…	…	…

figure, arrows are used to represent the adjustment directions of stocks (from H_r to h_r for retailer r). Seventeen retailers decrease their stocks, and also seventeen retailers increase their stocks by transshipment, whereas nine retailers do not change their stocks. Table 9.7 presents the transshipment amount for each pair of retailers, whereas Fig. 9.9 depicts the stock variance and boundaries for each retailer. The arrow presents the variance direction that means increase or decrease, whereas the boundaries restrict the lower and upper limits of the final stock. As shown in Fig. 9.9, the origin of an arrow is H_r (while circle) and the destination of the arrow is h_r (gray circle). The range of h_r is restricted with the lower and upper boundaries (the rectangle with a white and black box). This tool mainly presents the effects of transshipment on the stocks.

2. **Comparison of the solutions with or without transshipment**

The solutions with or without transshipment are compared in Table 9.8 by solving [M1] and [M2]. By using transshipment as an option, the total profit (f) is increased by 63.12 % and f^P is increased by 37.95 %, whereas the penalty costs of shortage or excess are all decreased considerably, by 91.04 and 88.11 %, respectively. Moreover, the transshipment just brings an increase of 2614.70 for the transportation cost. From the results of [M1] and [M2], transshipment contributes much to the penalty costs and the sales profit. As a summary, transshipment contributes to a significant profit improvement with a small increase of transportation cost. When the processes related to transshipment can be controlled, it is a practical way to decrease the in-store inventory.

3. **Sensitivity analysis for the parameters**

The performance of [M1] and [M2] depends on five groups of parameters, namely H_r, C^-, C^+, C^T, and C^P. Figures 9.10, 9.11, 9.12, 9.13, and 9.14 present their effects on the solutions, respectively.

Table 9.6 Stock and the boundaries of estimated demand for each retailer

No.	H_r	$\underline{V}$	$\overline{V}$
1	36	27	29
2	9	1	6
3	30	3	11
4	43	21	24
5	10	19	23
6	23	48	49
7	43	42	47
8	11	8	16
9	46	39	49
10	17	38	40
11	3	17	23
12	42	12	17
13	39	12	20
14	28	23	28
15	7	15	20
16	36	1	10
17	31	16	19
18	39	28	36
19	32	6	12
20	41	42	46
21	1	18	21
22	38	23	25
23	41	30	35
24	17	20	26
25	7	31	38
26	1	16	21
27	4	47	49
28	15	39	45
29	48	18	23
30	43	31	36
31	2	2	8
32	18	41	47
33	9	20	28
34	38	34	41
35	0	4	6
36	40	28	37
37	46	4	8
38	41	37	45
39	4	12	18
40	37	41	43
41	23	31	39
42	14	35	43
43	21	31	33

Table 9.7 Solutions of transshipment

r	e	$x_{r,e}$
1	5	7
1	6	2
2	5	2
2	10	5
3	10	8
3	11	16
4	6	22
9	10	3
12	20	2
12	21	17
12	43	10
13	24	5
13	26	2
13	27	17
14	27	4
16	15	9
16	25	12
16	33	11
17	25	5
17	32	10
18	10	5
18	33	3
19	26	4
19	31	2
19	42	18
22	32	15
23	40	4
23	42	6
29	26	8
29	39	10
29	41	11
30	25	9
30	26	2
34	35	2
36	27	7
36	35	2
37	27	15
37	28	26
38	5	1

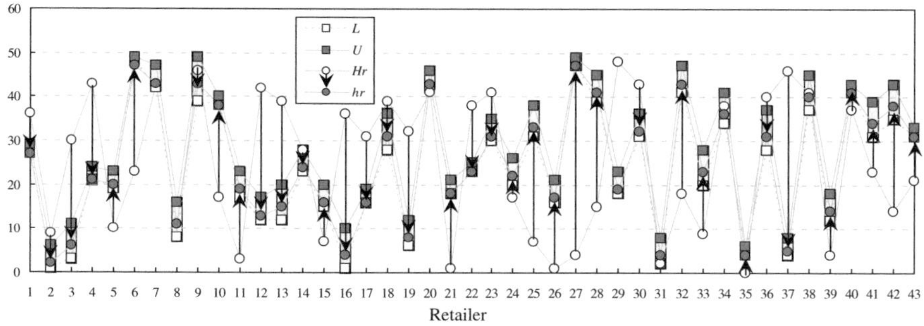

Fig. 9.9 The estimated sales and stock change for each retailer

Table 9.8 Comparisons of results by the solutions with or without transshipment

	f	f^P	f^T	f^+	f^-
No transshipment [M2]	314552.15	378715.82	0.00	31656.84	32506.84
Transshipment [M1]	513140.20	522432.07	2614.70	3763.59	2913.59
Increase	63.13 %	37.95 %	+2614.70	-88.11 %	−91.04 %

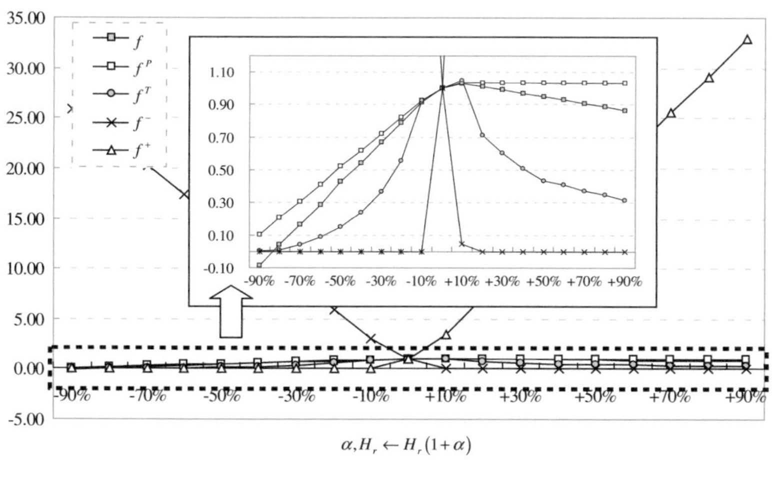

Fig. 9.10 Effects of the stock on solutions

Figure 9.10 presents the sensitivities of H_r. By changing H_r by a percentage (α from −90 to 90 %), the solution values of f, f^P, f^T, f^-, and f^+ are recorded. From Fig. 9.10 (the background figure), small changes can be seen for f, f^P and f^T, whereas f^- and f^+ vary much more. When α ranges from −90 to +10 %, f^- drops gradually to almost zero. When α ranges from −10 to +90 %, f^+ increases

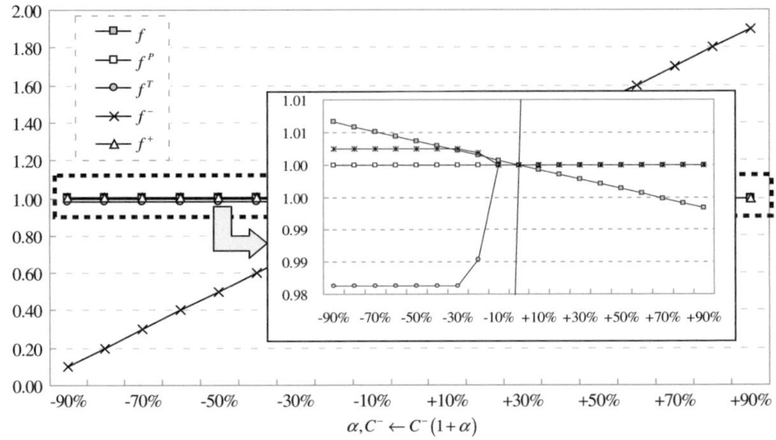

Fig. 9.11 Effects of the unit penalty cost of stock shortage on solutions

gradually from almost zero to a big amount. In order to see the detailed differences among f, f^P, and f^T, a new figure is used as the foreground in Fig. 9.10. As presented in this figure, f^P gradually increases and then stays stable after α reaches +30 %; f^T increases and then drops at $\alpha = 30\,\%$. f shares a similar varying tendency with f^P before α reaches +30 % and then drops slightly.

Figure 9.11 presents the effects of the unit penalty cost of stock shortage (C^-) on solutions. As shown in the background figure, f^+ almost does not change, whereas f^- linearly increases. From the foreground figure, f linearly decreases and f^P is almost unchanged; f^- keeps stable first and drops to one when α changes from -30 to 0 % and then keeps stable; f^T increases from its minimal to its maximal when α changes from -30 to 0 % and keeps stable at other values of α.

Figure 9.12 presents the similar effects of the unit penalty cost of stock excess (C^+) on solutions comparing to C^-. The distinct difference is f^- almost does not change, whereas f^+ linearly increases.

Figure 9.13 presents the complex effects of the unit cost of transshipment (C^T) on solutions. The changes of C^T impose a great effect on f^T, which gradually increases, as well as f^- and f^+. However, f^P is almost unchanged and f slightly and gradually decreases.

Figure 9.14 presents the effects of the unit profit of apparel (C^P) on solutions. Directly, f^P and f are affected most and increase with C^P. Although f^+, f^-, and f^T seem to be affected much, the percentages of variances indeed are small.

As a summary, the present stocks impose great effects on the two penalty costs and the transportation cost (Fig. 9.10). The penalty costs are greatly affected by the units of penalty costs (Figs. 9.11 and 9.12). The transportation cost is also mainly affected by the unit transshipment cost (Fig. 9.13). From Figs. 9.10, 9.11, 9.12, 9.13, and 9.14, although the parameters which have the greatest influence could be identified easily, it is necessary to analyze their cross-effects by the sensitivity

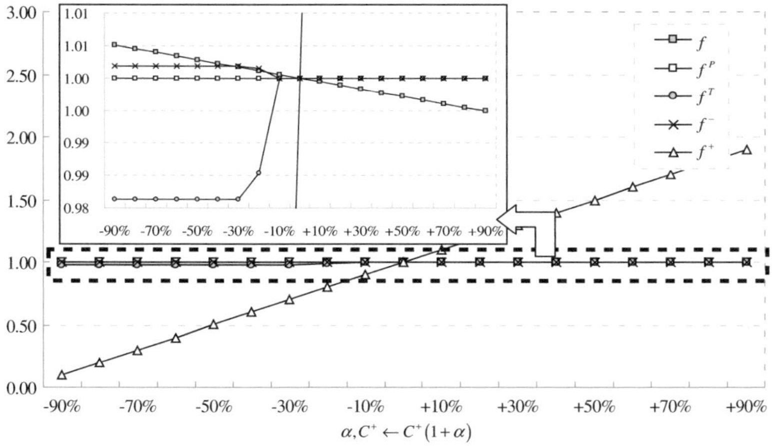

Fig. 9.12 Effects of the unit penalty cost of stock excess on solutions

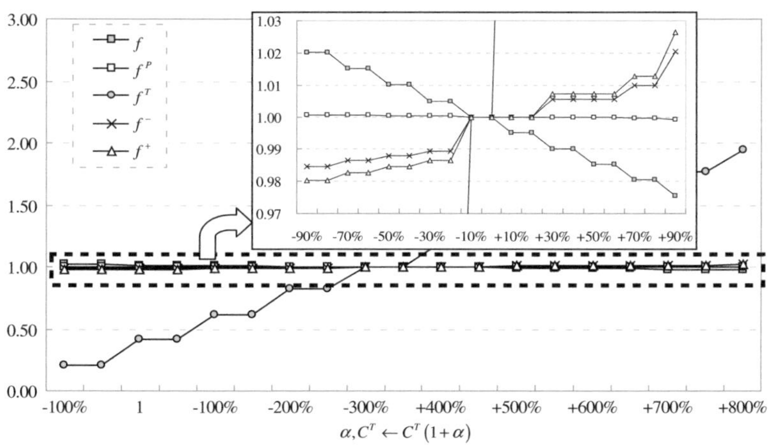

Fig. 9.13 Effects of the unit cost of transshipment on solutions

analysis tool. As presented in the small foreground figures in Figs. 9.10, 9.11, 9.12, 9.13, and 9.14, although the sensitivities are limited within a narrow scope (e.g., from 0.97 to 1.03), the effects on the solutions are apparent.

According to the results in Tables 9.4 and 9.5 and Figs. 9.9, 9.10, 9.11, 9.12, 9.13, and 9.14, transshipment is an effective way to reduce the penalty costs for stock shortage or excess, such that the total profit can be improved. Second, although the costs and profits of the solutions are directly affected by the corresponding parameters, the synthesis effects should be determined by sophisticated sensitivity analysis of the parameters. Third, this work mainly examines the

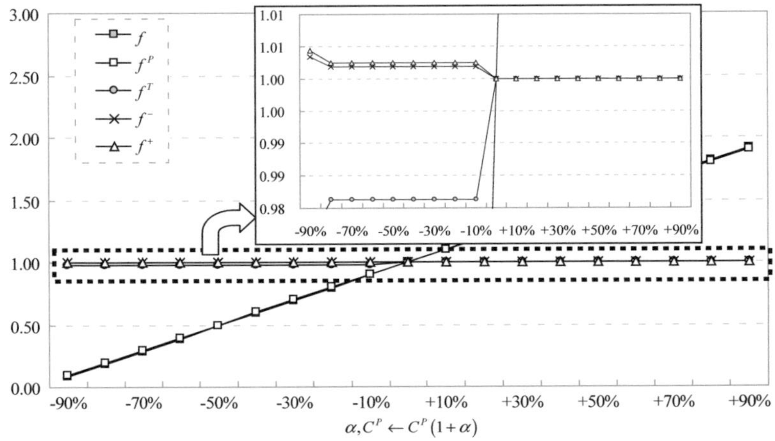

Fig. 9.14 Effects of the unit profit of apparel on solutions

decision support models because the effects of the parameters on the solutions depend on the relative quantitative relationships among them.

9.3.2.3 Discussions on Transshipment

This paper identifies transshipment as a strategy for sharing and reducing stocks for fashion apparel. Although the fast fashion formula has been widely studied in the scopes of supply chain management, marketing and customer behavior management, and fashion retailing, in particular, transshipment as a simple logistics strategy that quickens the turnover of stocks deserves much deeper research. The fashion apparel transshipment problem has the following features. First, the transshipments among retailers are coordinated such that an alliance is introduced to manage the comprehensive transshipments among retailers. Second, the sales are estimated and treated as stochastic numbers in the proposed model. The resulting formulation is then a stochastic programming model. Because transshipment is used to share stocks, the available stocks and the sales in the next periods are then all uncertain. Third, the transshipment decision depends on many other decision processes. Therefore, we propose a DSS to include it for improving the applicability. Based on this work, the replenishment and profit sharing mechanisms among retailers can be further studied in order to improve the applicability. However, these studies will make the integrated problem more complex for modeling and analysis. We would postpone it to future research.

9.3.3 Sustainable Operations

In the fashion industry, the used clothing may be resold, reused, and donated, and most of them are finally discarded (Bianchi and Birtwistle 2012). Textiles in municipal solid waste are found mainly in discarded clothing, although other sources include furniture, carpets, tires, footwear, and non-durable goods such as sheets and towels. Moreover, in many long processes of clothing manufacturing, the environment is heavily polluted and many resources are wasted. Textile recycling is the method of reusing or reprocessing used clothing, fibrous materials, and clothing scraps from the manufacturing process. As an intuitive and sustainable strategy, reusing the clothing can reduce the manufacturing pollution and resource consumption. However, consumers have different opinions and behaviors on clothing disposals, recycling, and remanufacturing.

Based on the features of fast fashion products, this work develops a rent-based system for improving the level of sustainability. Fast fashion products usually have shorter life cycles than regular "non-fast fashion" clothing products; the costs of these products are relatively higher; the fashion product at a city may be transferred to other cities in a fashion cycle. When the fashion products are consumed by rental services, these features can be utilized to improve the level of sustainability. Based on these considerations, this work develops a sustainable rent-based supply chain for fashion products. The sustainability and the strategies to improve the sustainability of the supply chain are investigated. However, our developed supply chain system is workable on the premise that the consumers can accept the fashion products that are rented cyclically. This premise should be inspected in real life by further empirical studies. In this work, we consider the rent-based system because first, financial and equipment leasing operations have been well established in the real world. Second, many products, including furniture, computers, and other products that can be temporarily used, also have been rented. Third, many special types of clothing products have been supported by rental markets. In China, wedding apparels, performance costumes, and other formal or informal dresses have been widely leased by many specialized fashion companies.

Comparing to the studies on fashion and textile industry, supply chain design, and sustainable fashion products, this study contributes to the literature in the following aspects. First, two closed-loop supply chain models are proposed for fashion products in the regular retailing channel mode and the rent-based mode, respectively. The rental service involves the primary contribution, and all schemes are developed by considering the business mode based on the presence of rental service. Second, the activities in the rent-based sustainable supply chain are compared with the regular retailing channel mode, based on some strategies that have been researched for sustainable fashion retailing (Choi and Chiu 2012). Moreover, the sustainable operational strategies are devised for these activities. Further, the managerial implications for improving the sustainability are examined and discussed. The key feature of this work involves a system of rental-based sustainable supply chain and a framework of research aspects. Based on them, special business modes, consumer behaviors, and sustainability values can be further examined. Practitioners can use the research results to enhance their operations.

9.3.3.1 The Sustainable Fashion Supply Chain

Considering fashionable clothing products, a system of sustainable rent-based closed-loop supply chain is presented in Fig. 9.15. In the figure, eight types of stakeholders and four important processes are presented.

The rent-based supply chain is described as follows. The fashion products used by rental services are manufactured according to the designs. The designs must be based on the careful investigations of the markets. Specific details of the fashion system include garment sizes, materials, colors, and patterns, which should be perfectly examined by using data from the target market. Some consumers may become "fashion experts" after they experience the buying and rental services. They may have different judgments on the fashion trends and consumer perceptions. The products are distributed to consumers by two channels, namely the traditional retail

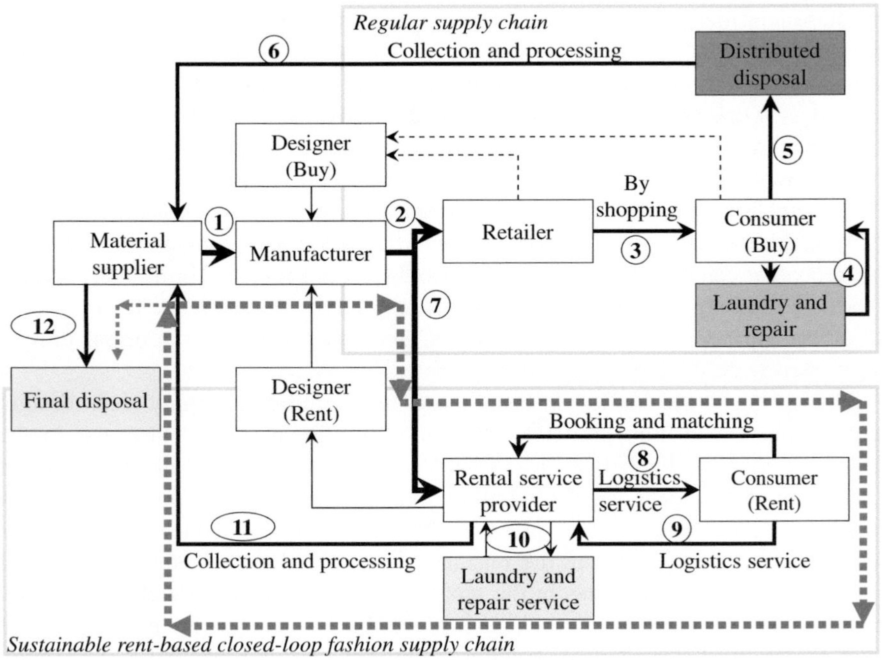

Fig. 9.15 A system of sustainable rent-based supply chain. (*Note* the *white boxes* represent the primary stakeholders; the *gray boxes* are important processes, and the texts on the *arrows* also indicate processes; the *dark bold colored arrows* indicate transportation activities; the thin (*dotted* or *not*) *arrows* indicate information flows; the *gray dotted arrows* construct a closed loop of fashion products.)

sales channel and the rental services channel. From the traditional retail sales channel, the consumers buy fashion products from the retailers and wash the products in use or dispose the products by themselves in a distributed manner. Some of the disposals may be collected and reused by the material suppliers. Through the rental service channel, consumers book the fashion products by either an online or an off-line approach. The matched products are used by the consumers and then sent back to the rental service providers by employing logistics services. The laundry and repair processes are handled by the rental service providers themselves or outsourced. Anyway, the products will be collected and treated in batches periodically. The laundry and repair process includes washing, disinfection, and repair. Most products will be used again for other customers under the rental services. Some out-of-date or worn-out products will be sent to the material suppliers for further reuse of engineering processes. By retailing and especially rental services, the data of consumer buying and rental behaviors can be collected for forecasting the fashion trends. Actually, the rental service would have a strong fashion discovery capability than pure retailing because the consumers may communicate with the rental service providers with a higher frequency.

Closed-loop supply chain and logistics systems are well studied in literature (Bassamboo et al. 2009; Wang and Hsu 2010). Notably, the term "closed loop" means a high degree of utilizing the resources in a loop. In this work, we study the sustainable rent-based supply chain by introducing a conceptual model and developing operations management strategies by reviewing the related literature. The sustainability of the supply chain is examined in the following aspects: resource consumption, labor and power usage, operations scales, and difficulties.

1. **Sustainable laundry**

Water is used in large quantities in (distributed) laundries, which at present produce relatively high quantities of wastewater (Šostar-Turk et al. 2005), in the retailing channel in Fig. 9.15. In the rental services, general laundry procedures can be more effective, sustainable, and environmentally friendly. Conclusively, the business mode based on the rental service can generate many opportunities to improve the sustainability of laundry operations in fashion supply chains by controlling the operations, chemicals, recycling, and reusing of water and saving energy.

2. **Green logistics**

With respect to environment pollution, transportation is the most visible aspect of supply chain. CO2 emissions of transportation occupy about 14 % of total emissions, both at global and at the EU level (Dekker et al. 2012). Transportation is also a main source of NOx, SO2, and PM (particulate matter or fine dust) emissions. The storage of products also has an environmental impact. Inventory holding costs play a large part in supply chain design, as the more storage is centralized, the less the storage costs. In the rent-based supply chain, more activities are scaled such that they can be more fully controlled to increase the sustainability. We may consider that the transportation activities 8 and 9 (Fig. 9.15) have negative impacts. Actually, these activities are usually outsourced to the third-party logistics (3PL) firms, which indeed help minimize the impacts. In other activities, the centralized and scaled processes of the rental service contribute to the savings of transport power, energy, and pollutions.

3. **Green disposal**

Textile disposal is an increasing problem in the world. Fast fashion retailing is leading consumers toward an increased rate of purchasing which may result in more clothing disposal. Bianchi and Birtwistle (2012) empirically explored two methods of sustainable clothing disposal behaviors: donating to charities and giving away to family and friends. Their results show that the consumer recycling behavior is a strong and direct driver leading consumers to donate used apparel for charity. In addition, consumer awareness of the environment protection and consumer age affects the donating behaviors. The influences of attitudes toward the environment and subjective norms of family and friends on clothing disposal behaviors were also examined in Bianchi and Birtwistle (2012). Their results indicated that resale and donation behaviors were explained by environmental concerns and that reuse and resale behaviors were explained by economic concerns. Five disposal options in the supply chains are resale, donation, reusing, recycling, and discarding (Joung and Park-Poaps 2013). In the regular supply chain, the used fashion products may be disposed by donation, resale, and discarding. The diverse consumers' behaviors play important roles in the disposal. In developing countries, when the donation bins and charity shops are not well set, the convenience concerns may result in more undesirable disposal behaviors. However, the rent-based supply chain instinctively serves as a scheme for reusing the products, thus enhancing sustainability.

4. **Sustainable design**

Theoretically, our work demonstrates a systematic and logistical procedure for the identification of energy factors for sustainable fashion and the development of feasible and practical scenarios for the design and production. The short life span of textiles and especially fashion products is one of the main problems in the current industrial system based on planned obsolescence. Three strategies are discussed here. The first strategy is about long lifetime guarantee and product satisfaction. The life span of the product and the quality of textiles and garments are difficult to evaluate at the point of purchasing. Therefore, maintenance quality is critical. At this point, the rental service is a better choice. Secondly, designs of products and their attachments should be "emotionally satisfying." By rental services, the preference and emotion can feedback efficiently. A few typical processes are involved: customization, semifinished products, and modular structures; cocreation; and open source design. The third strategy relates to services. Product–service systems emphasize systems thinking and drive companies to focus on consumer needs. Longer product life spans can also be achieved through services such as upgrading or updating, repairing, or product modification systems or services. These services extend the enjoyable time of the fashion product as well as postpone the psychological obsolescence that consumers themselves feel about the product.

5. **Sustainable manufacturing**

In the context of the fashion product life cycle, the concept of sustainability can be seen in terms of three stages: manufacturing, utilization, and disposal. Issues such as supply of raw materials, production of final products, application of chemical treatments, and/or processes of operational and logistical activities are involved in the manufacturing stage. In the fashion supply chain depicted in Fig. 9.15, comparing with the other processes, the rental service imposes little effect on the manufacturing stage. However, the rental service may change the scale of manufacturing and the consumption modes. In the manufacturing scale, the reusing and recycling scales are enlarged. Through frequent and efficient face-to-face sales and logistics services, sustainable materials, designing, and manufacturing technologies can be reported to the consumers, and thus, the rental service can guide their sustainable consumption. The utilization and disposal stages are different because of increased scales in the rent-based supply chain (Fig. 9.15).

9.3.3.2 Operations Management Issues

1. **Centralized versus distributed inventory management**

In Fig. 9.16, the centralized and distributed inventory flows of fashion products are presented. Notably, the disposal flows are not included in the figure. In the centralized one (Fig. 9.16a), the products are distributed from the manufacturer to the warehouses. Each warehouse should serve at least one rental service provider. But, only some typical samples are distributed from the warehouse to the rental service providers. After the consumers determine their rental plans by trying the samples with the help of the servicers, the warehouse will distribute the product to the consumers by 3PL. The used products after the rental period will be collected by 3PL from the consumers to the warehouse. In Fig. 9.16, only the warehouses manage the big inventories. The distributed one (Fig. 9.16b) depicts more complex logistics flows. The products are generally not distributed from warehouses to consumers, but through the rental service providers to consumers. Therefore, rental service providers also manage rather big inventories. Further, to share the inventory, logistics transshipment may occur among the warehouses and rental service providers. Comparing the two modes, the following differences are observed. In the centralized one, the inventories are centrally managed such that they can be efficiently utilized; consumers cannot take products home directly after ordering them at the rental service providers; small product packages are directly distributed to or collected from consumers by 3PL, and thus, seemly more transportation costs, energy, and impacts on environment are involved. In the distributed one, the consumer can experience at the rental sites and the satisfaction may be improved,

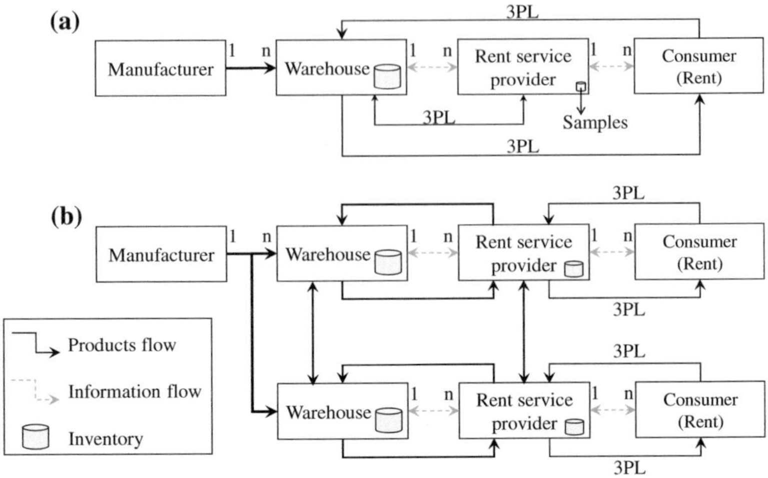

Fig. 9.16 The management modes of centralized and distributed inventory of fashion products. **a** Centralized inventory of fashion products. **b** Distributed inventory of fashion products

whereas the online booking system does not yield this kind of satisfaction; scaled transportation between warehouses and rental service providers can be performed; however, the inventory management becomes complex.

2. **Sustainable logistics services**

Two levels of logistics are also presented in this study: two big closed-loop logistics (one for the regular fashion supply chain and another for the rent-based closed-loop fashion supply chain) and a small but frequently used internal closed-loop logistics in the rent-based supply chain. In the top closed-loop logistics (Fig. 9.17), some detailed transportation and transshipment links are not depicted. In this loop, the forward flows are products, whereas the reverse flows are used products. Because the used products come from the rental service providers after cleaning and proper treatment, they are safe and can be transported together by containers with the forward flows. An exception is the flow to the final disposal. Nevertheless, this kind of flow usually can be compacted and efficiently transported. The internal closed-loop logistics when performed by a vehicle fleet can be formulated as a vehicle routing problem with pickup and delivery. The vehicle spaces can be utilized by the forward and reverse product flows simultaneously. Based on the above descriptions and analysis, the logistics can be put into sustainable closed-loop logistics practices.

3. **Human–clothing matching methods**

Inspired by the literature (Cho et al. 2005; Hu et al. 2013; Ding et al. 2011) and their research findings considering human–clothing matching problems, the features of rent-based fashion supply chain are examined, and thus, the possible applications

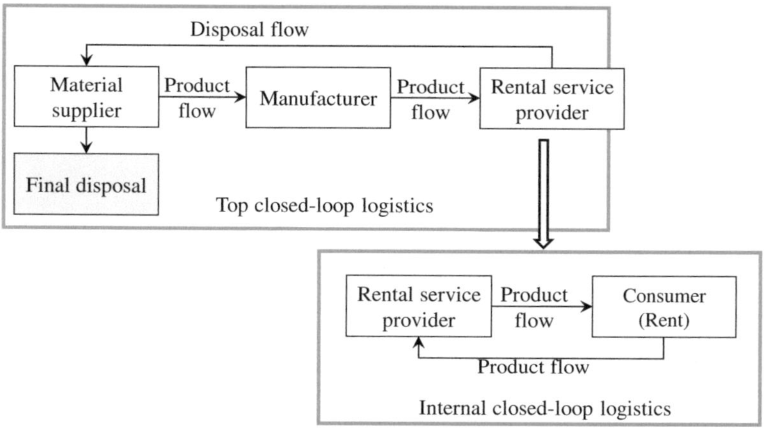

Fig. 9.17 A system of sustainable rent-based supply chain

Table 9.9 Inspirations of human–clothing matching methods for the rent-based supply chain

Literature	Inspirations
Cho et al. (2005)	1. Interactively measure consumer body sizes and determine fit fashion products 2. Establish the detailed human body measurement model for customization and adjustments
Hu et al. (2013)	1. Build the batch allocation method for large-scale rental tasks 2. Build the scheduling system based on the multisize matching model
Ding et al. (2011)	1. Use the MSCD model with size satisfaction for fashion product matching; 2. Extend the idea for building matching tool for online booking system

and extensions are proposed in the table. The inspirations in the last column (Table 9.9) are presented for the purpose of improving the rental service quality by reducing the try-on times. These inspirations are also effective for general apparel retailing practices. The human–clothing matching models help to improve the efficacy of rental services when many fashion products are available for the consumers.

4. **Booking systems under uncertainty**

Booking as a strategy coping with uncertainty is usually coupled with revenue management. Revenue management enhances the revenues of a company by means of demand management decisions. The booking system in rental services can also be used to diminish the effects of demand uncertainty on the utilization ratios of products. In particular, driven by the fast fashion trend, the turnover ratio or the rent times are important indices for improving the profitability of fashion products. Figure 9.18 presents the information and product flows in the booking system. By

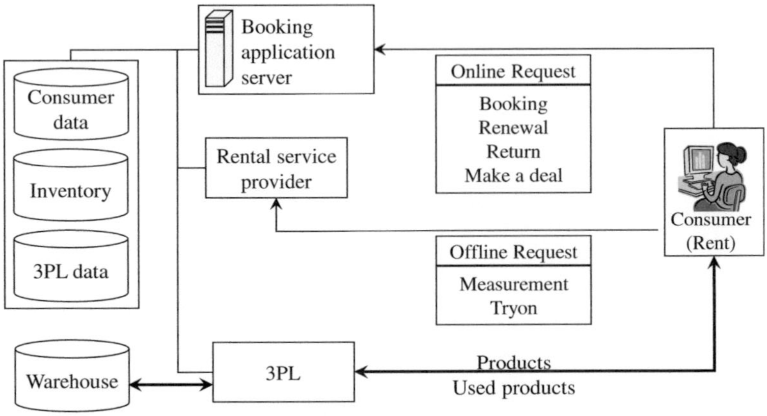

Fig. 9.18 A typical booking system

the booking system, a consumer can submit four types of online requests to the rental service provider: booking, renewing a booking order, return of used products, and making a deal by e-commerce. Initially, a consumer may not complete the booking request by the online platform. She or he should go to the rental service provider to measure her or his body and have some trial of the products. These off-line activities will generate the consumer's body data that may be repeatedly used in the future booking request. The booking uncertainty can be mediated by the pricing strategy that is elucidated in the next subsection.

5. **The competitive rental pricing problem**

Pricing is well studied in the context of supply chain management. The competition of the rental pricing problem has the following features. First, the rental products should be competitive to the new products on sale. However, because the rental products generally are used for a short term, the competitive advantages should be further measured. Second, the rental products should not intensify the price competition with the products that are regularly sold. Rental service and retailing aim at different markets. Third, in the rental pricing problem, the sales and logistics services, as well as the green, environment, and social values of using sustainable fashion products, should be incorporated into the model. These features should be considered in the pricing models by extending the prior studies on pricing.

9.3.3.3 Discussions on Sustainable Operations

In addition to the operations management issues investigated above, a more primitive problem is proposed, and why would the market accept the sustainable rental fashion products and the corresponding consumption style of wearing? Some sustainability promotion strategies are discussed below to enhance the market acceptance to the fashion product rental services:

1. Fashion customization and design: Three important measures are proposed to speed up the fashion customization design, namely the interactive codesign scheme with consumers, implementing mass customization by new technologies and optimization methods and achieving fast redistribution among regions.
2. Responsive system for fashionable products: Fashion itself is full of uncertainty. To cope with this fashion feature, a practical way is to respond to the fashion trend as fast and efficiently as possible. Four aspects for fast response include information system integration (among various stakeholders), booking requirement management, consumer management (e.g., according to the consumers' sizes, the concise and an accurate human–clothing matching can filter the candidate fashion products for recommendation), and fast logistics.

9.4 Challenges and Opportunities of Using Advanced Technologies

9.4.1 O2O-Based Fashion Logistics Systems

The rapid development of information technology enables an increasing number of consumers to search and book products and services online first and then to buy and pick up them in brick-and-mortar stores. This new e-commerce model is called online to off-line (O2O) e-commerce and has received significant attention in recent years. In the literature, Liao et al. (2015) considered a hotel distribution system in which a hotel sells rooms through both its own off-line channel and the "online travel agency's (OTA)" online channel. In particular, the authors considered a commonly used allocation scheme in the hotel industry: the "commission override model (COM)," which uses both the wholesale pricing contract and the consignment contract to sell hotel rooms. Compared with many extant e-commerce models (i.e., B2B, B2C, and C2C), reputation management in this emerging model needs some improvement. It has to collect more raw reputation-related data, consider more reputation-related factors, and show more comprehensive reputation evaluation results. As a stepping stone in the research in O2O e-commerce, a new reputation management system (HSMM-RMS) has been developed based on a probabilistic model called the hidden semi-Markov model. By combining the observable online and off-line raw reputation data, the proposed system can accurately, promptly, and dynamically provide O2O e-commerce participants with off-line merchants' historical and predictive reputation information. Xiao and Dong (2015) indicated that the proposed HSMM-RMS system performs significantly better than the extant hidden Markov model-based reputation management system by Monte Carlo simulation experiments. Xiao and Dong (2015) further conducted a case study based on a real O2O e-commerce platform to demonstrate the application of HSMM-RMS.

Data mining (DM) techniques have been used to solve marketing and manufacturing problems in the fashion industry (Brito et al. 2015) and can be extended for exploring O2O business models when the transactions and consumer data can be collected and utilized. These approaches are expected to be particularly important for highly customized industries because the high diversity of products sold makes it difficult to find clear patterns of customer preferences. Brito et al. (2015) investigated two different data mining approaches for customer segmentation: clustering and subgroup discovery. Their models produced six market segments and 49 rules that allowed a better understanding of customer preferences in a highly customized fashion manufacturer/e-tailor.

9.4.2 IoT-Based Fashion Logistics

With the development of the Internet of Things (IoT) technology, the logistics information management achieved by the IoT and related RFID technology and GPS can solve many logistics problems. In the related literature, Sun (2012) designed a kind of cargo transportation management system combining RFID, GPS, and GPRS technologies in which RFID is used to record (into) the goods information; GPS is for the orientation of the vehicle; and GPRS helps to transfer the corresponding data information. In order to study the effect of the Internet of Things (IOT) on the architecture of a logistics service supply chain (LSSC), Liu and Gao (2014) summarized the application of IOT in related fields on the basis of the theories of IOT and analyzed the effect of IOT on logistics/service flows, information flows, and fund flows in a LSSC and the effect on the structure of a LSSC. They established the architecture of the LSSC based on IOT and finally forecasted the application prospect of IOT in LSSC management. In view of the lack of efficiency of the automatic logistics management in China, Gao (2012) introduced the IoT technology to help realize the tracking and confirmation of logistics products. Undoubtedly, IoT can be used as a solid technology for improving fashion product logistics and supply chain management.

9.4.3 Experience-Based Fashion Logistics

Logistics services are now accessed very frequently and support e-commerce models. However, through logistics services (such as direct delivery), the consumers may fail to experience some important services from fashion sellers (such as personal fashion advice and sharing). Schmitt et al. (2015) presented a consumer experience model that views materialism and experientialism as two separate dimensions which affect consumer happiness (both in the form of pleasure and in the form of meaning), depending on the type of brand experiences being evoked. Thus, a good life in a consumerist society means integrating material and experiential consumptions rather than shifting spending from material to experiential purchases. Hu and Jasper (2015) examined the role that consumers' shopping experience plays when consumers choose online or shopping malls to shop. Therefore, it is beneficial to create new modes for incorporating shopping experience into logistics services. For example, O2O is an option to provide off-line consumer experiencing.

9.5 Conclusions

In this chapter, fashion logistics management has been investigated by analytical analysis and numerical experiments in distribution, transshipment, and sustainable logistics. A summary of results is presented below.

1. Try-on service is considered during the distribution process. The try-on time uncertainty is coped with by three strategies: adjusting the try-on time estimation parameters, adjusting the time reservation ratio for each route, and applying the try-on service rejection when the distribution time is insufficient for the remaining customers in the route. An evolutionary algorithm is devised and implemented for optimizing routes with the minimum logistics cost and simulating uncertain try-on times.
2. Retailing coupled with transshipment as a simple logistics strategy quickens the turnover of stocks. The fashion apparel transshipment problem considers the following features. The transshipments among retailers are coordinated such that an alliance is introduced to manage the comprehensive transshipments among retailers. The sales are estimated and treated as stochastic numbers in the proposed model. The transshipment decision depends on the decision processes.
3. For sustainable logistics management, the fashion product rental service is explored which may improve the level of sustainability in three levels, namely the processes, the operations, and the promotion levels.

Finally, we have also discussed some probable new research directions, e.g., by employing latest information technologies such as IoT and the O2O business model.

As a remark, in the future, to solve complex logistics management problems like the ones in transshipment, we need to explore new optimization methods and techniques. Last but not least, we should pay attention to the fact that it is important, though challenging, to extend the analysis for a very comprehensive fashion logistics system in which the topics (distribution, transshipment, and sustainable logistics) are considered together. This will involve new optimization objectives and require more advanced and efficient new methods in order to find the optimal solutions.

References

Agrawal S, Singh RK, Murtaza Q (2015) A literature review and perspectives in reverse logistics. Resour Conserv Recycl 97(1):76–92
Barnes L, Lea-Greenwood G (2006a) Fast fashioning the supply chain: shaping the research agenda. J Fashion Mark Manage 10(3):259–271
Barnes L, Lea-Greenwood G (2006b) Guest editorial. Journal of Fashion Marketing and Management 10(3):1
Barnes L, Lea-Greenwood G (2010) Fast fashion in the retail store environment. Int J Retail Distrib Manage 38(10):760–772
Bassamboo A, Kumar S, Randhawa RS (2009) Dynamics of new product introduction in closed rental systems. Oper Res 57(6):1347–1359
Bianchi C, Birtwistle G (2012) Consumer clothing disposal behaviour: a comparative study. Int J Consum Stud 36(3):335–341
Bing X, Bloemhof-Ruwaard JM, Van Der Vorst JGAJ (2014) Sustainable reverse logistics network design for household plastic waste. Flex Serv Manuf J 26(1–2):119–142
Bouhouras E, Basbas S (2015) Policies towards sustainable city logistics. The case of Thessaloniki. J Environ Prot Ecol 16(2):417–423
Brito PQ, Soare C, Almeida S, Monte A, Byvoet M (2015) Customer segmentation in a large database of an online customized fashion business. Robot Comput-Integr Manuf 36(1):93–100
Caniato F, Caridi M, Crippa L, Moretto A (2012) Environmental sustainability in fashion supply chains: an exploratory case based research. Int J Prod Econ 135(2):659–670
Cheng Z, Xiao J, Xie K, Huang X (2013) Optimal product quality of supply chain based on information traceability in fashion and textiles industry: an adverse logistics perspective. Math Prob Eng Article ID 629363. http://dx.doi.org/629310.621155/622013/629363
China Internet Network Information Center (2015) (Retrieved on 11 Dec 2015): URL: http://www.cnnic.net.cn/hlwfzyj/hlwxzbg/dzswbg/201509/P020150909354828731159.pdf
Cho Y, Okada N, Park H, Takatera M, Inui S, Shimizu Y (2005) An interactive body model for individual pattern making. Int J Clothing Sci Technol 17(2):91–99
Choi T-M, Chiu C-H (2012) Mean-downside-risk and mean-variance newsvendor models: Implications for sustainable fashion retailing. Int J Prod Econ 135(2):552–560
Choi TM, Chiu CH, To KMC (2011) A fast fashion safety-first inventory model. Text Res J 81 (8):819–826
Choi T-M, Yue X, Chiu C-H, Chow P-S (2013) Mathematical modeling research in fashion and textiles supply chains and operational control systems. Math Prob Eng Article ID 470567. http://dx.doi.org/470510.471155/472013/470567
Dekker R, Bloemhof J, Mallidis I (2012) Operations research for green logistics—An overview of aspects, issues, contributions and challenges. Eur J Oper Res 219(3):671–679
Digiesi S, Mossa G, Mummolo G (2012) A loss factor based approach for sustainable logistics. Prod Plann Control 23(2–3):160–170
Ding Y-S, Hu Z-H, Zhang W-B (2011) Multi-criteria decision making approach based on immune co-evolutionary algorithm with application to garment matching problem. Expert Syst Appl 38 (8):10377–10388
Dong Y, Xu K, Evers PT (2012) Transshipment incentive contracts in a multi-level supply chain. Eur J Oper Res 223(2):430–440
Eksioglu B, Vural AV, Reisman A (2009) The vehicle routing problem: a taxonomic review. Comput Ind Eng 57(4):1472–1483
Eliiyi D, Z YE, Şahin DY (2011) Supply chain management in apparel industry: a transshipment problem with time constraints. J Text Apparel 21(2):176–181
Eskandarpour M, Dejax P, Miemczyk J, Péton O (2015) Sustainable supply chain network design: An optimization-oriented review. Omega (United Kingdom) 54(1):11–32
Fahimnia B, Sarkis J, Davarzani H (2015) Green supply chain management: a review and bibliometric analysis. Int J Prod Econ 162(1):101–114

Gao LZ (2012) Establishment of smart logistics management system based on IOT technology. Logistics Technol 12(1):23–26

García-Arca J, Carlos Prado-Prado J, Trinidad Gonzalez-Portela Garrido A (2014) "Packaging logistics": promoting sustainable efficiency in supply chains. J Phys Distrib Logistics Manage 44(4):325–346

Gong Y, Yücesan E (2012) Stochastic optimization for transshipment problems with positive replenishment lead times. Int J Prod Econ 135(1):61–72

Grahovac J, Chakravarty A (2001) Sharing and lateral transshipment of inventory in a supply chain with expensive low-demand items. Manage Sci 47(4):579–594

Herer YT, Rashit A (1999) Lateral stock transshipments in a two-location inventory system with fixed and joint replenishment costs. Naval Res Logistics 46(5):525–547

Hines T, Bruce M (2001) Fashion marketing: contemporary issues. Butterworth-Heinemann, Oxford

Hu H, Jasper CR (2015) The impact of consumer shopping experience on consumer channel decision. Acad Mark Stud J 19(1):213–224

Hu Z-H, Yu X-K, Sheng Z-H (2013) An immune co-evolutionary algorithm based approach for problems related to clothing uniform assignment. Int J Clothing Sci Technol 25(1):70–82

Joung H-M, Park-Poaps H (2013) Factors motivating and influencing clothing disposal behaviours. Int J Consum Stud 37(1):105–111

Kim G, Ong Y-S, Heng CK, Tan PS, Zhang NA (2015) City vehicle routing problem (City VRP): a review. IEEE Trans Intell Transp Syst 16(4):1654–1666

Liao P, Ye F, Tayi GK, Zhao X (2015) Competition and coordination in online and offline hotel distribution channels under commission override model. Int J Revenue Manage 8(2):193–217

Liu W, Gao Z (2014) Study on IOT based Architecture of logistics service supply chain. Int J Grid Distrib Comput 7(1):169–178

Liu S, Lei L, Park S (2008) On the multi-product packing-delivery problem with a fixed route. Transp Res Part E: Logistics Transp Rev 44(3):350–360

Macchion L, Moretto A, Caniato F, Caridi M, Danese P, Vinelli A (2015) Production and supply network strategies within the fashion industry. Int J Prod Econ 163:173–188

Naseraldin H, Herer YT (2011) A location-inventory model with lateral transshipments. Naval Res Logistics 58(5):437–456

Ngai EWT, Peng S, Alexander P, Moon KKL (2014) Decision support and intelligent systems in the textile and apparel supply chain: an academic review of research articles. Expert Syst Appl 41(1):81–91

Paterson C, Kiesmüller G, Teunter R, Glazebrook K (2011) Inventory models with lateral transshipments: a review. Eur J Oper Res 210(2):125–136

Pillac V, Gendreau M, Guéret C, Medaglia AL (2013) A review of dynamic vehicle routing problems. Eur J Oper Res 225(1):1–11

Potvin J-Y (2009) State-of-the art review: evolutionary algorithms for vehicle routing. INFORMS J Comput 21(4):518–548

Prins C (2004) A simple and effective evolutionary algorithm for the vehicle routing problem. Comput Oper Res 31(12):1985–2002

Ramos TRP, Gomes MI, Barbosa-Póvoa AP (2014) Planning a sustainable reverse logistics system: Balancing costs with environmental and social concerns. Omega (United Kingdom) 48 (1):60–74

Rosales CR, Rao US, Rogers DF (2013) Retailer transshipment versus central depot allocation for supply network design. Decis Sci 44(2):329–356

Ryvkin D (2010) The selection efficiency of tournaments. Eur J Oper Res 206(3):667–675

Schmitt B, Joško Brakus J, Zarantonello L (2015) From experiential psychology to consumer experience. J Consum Psychol 25(1):166–171

Šostar-Turk S, Petrinić I, Simonič M (2005) Laundry wastewater treatment using coagulation and membrane filtration. Resour Conserv Recycl 44(2):185–196

Stanger SHW, Wilding R, Hartmann E, Yates N, Cotton S (2013) Lateral transshipments: an institutional theory perspective. Int J Phys Distrib Logistics Manage 43(9):747–767

Sull D, Turconi S (2008) Fast fashion lessons. Bus Strategy Rev 19(2):4–11
Sun J (2012) Design and implementation of IOT-based logistics management system. In: IEEE Symposium on IEEE, Electrical & Electronics Engineering (EEESYM) 2012, vol 12, issue no. 12, pp 603–606
Ting C-K, Su C-H, Lee C-N (2010) Multi-parent extension of partially mapped crossover for combinatorial optimization problems. Expert Syst Appl 37(3):1879–1886
Tokatli N, Wrigley N, Kizilgün Ö (2008) Shifting global supply networks and fast fashion: made in Turkey for Marks and Spencer. Glob Netw 8(3):261–280
Touboulic A, Walker H (2015) Theories in sustainable supply chain management: a structured literature review. Int J Phys Distrib Logistics Manage 45(1):16–42
Turker D, Altuntas C (2014) Sustainable supply chain management in the fast fashion industry: An analysis of corporate reports. Eur Manag J 32(5):837–849
Wang H-F, Hsu H-W (2010) A closed-loop logistic model with a spanning-tree based genetic algorithm. Comput Oper Res 37(2):376–389
Xiao S, Dong M (2015) Hidden semi-Markov model-based reputation management system for online to offline (O2O) e-commerce markets. Decis Support Syst 77(1):87–99
Xifeng T, Ji Z, Peng X (2013) A multi-objective optimization model for sustainable logistics facility location. Transp Res Part D: Transport Environ 22(1):45–48

Chapter 10
Design of Order-Picking System and Selecting Picking Strategies in a 3PL Firm for Serving Fashion Retailing Companies

Aysenur Can and Feyzan Arikan

Abstract This study considers a real-life, picker-to-part order-picking system employing humans with multiple picks per route in a third-party logistics (3PL) firm. The 3PL firm has many customers from different sectors. The customer, whose orders are subject to this study, is primarily engaged in retail sales in the fashion industry. Being competitive in the fashion market relies on the ability to be flexible and responsive to meet the customer's demands. In the fashion business, lead time, delivery reliability, and stock availability are three key elements for satisfying the customer's perceived service level. The effective management of these three elements is closely linked to the process of distribution and logistics, which implies that the fashion firms should choose and cooperate with an efficient logistics service supplier. To maintain the efficiency of the 3PL firm that directly supports its fashion retailing customers, an order-picking system for the customer's fashion products is designed. To choose the most appropriate order-picking system, a process flow consisting of strategic decisions, operational implementation, and evaluation and selection (ES) sections is developed. In the operational level of the flow, the pick-and-sort batch-picking strategy with two different levels for the picker number and the synchronized zone-picking strategy with three different zone sizes are utilized as five alternative order-picking strategies based on the considered problem. For each alternative case, corresponding mathematical models of the order-batching problem are solved to maximize the total number of items fulfilled. In the ES section, the order-picking strategy with the lowest cost is selected based on the cost-centric analytical approach for alternative strategies.

A. Can (✉) · F. Arikan
Faculty of Engineering, Industrial Engineering Department,
Gazi University, 06570 Maltepe, Ankara, Turkey
e-mail: aaysenurcan@gmail.com

F. Arikan
e-mail: farikan@gazi.edu.tr

T.-M. Choi (ed.), *Analytical Modeling Research in Fashion Business*,
Springer Series in Fashion Business, DOI 10.1007/978-981-10-1014-9_10

Keywords Order picking · Warehouse · Distribution center · Design · Strategy selection · Application study

10.1 Introduction

A supply chain as an enhanced network of suppliers, manufacturers, warehouses, distribution centers, retail locations, or vendors and customers includes five major functions: research and development, purchasing, production, quality, and logistics chains. Among them, the logistics chain maintains the essential distribution network in a supply chain. The efficiency and effectiveness in any distribution network in turn are largely determined by the operation of the nodes in such a network, i.e., the warehouses (Rouwenhorst et al. 2000).

In the past, warehouses were seen mainly as stock-holding points, whereas today, they usually have many different roles including buffering the material flow along the supply chain to accommodate variability caused by factors such as product seasonality and/or batching in production and transportation, consolidation of products from various suppliers for combined delivery to customers, and value-added processing such as kitting, pricing, labeling, and product customization (Gu et al. 2007). All warehouse operations can be owned, leased, or operated by third-party companies on behalf of a principal. Warehouses operated by third-party logistics providers are either dedicated operations on behalf of a single customer or can be shared-user or public warehouses where a number of different customers share resources and are accommodated under one roof (Richards 2011).

Although warehouses differ in terms of their size, type, function, ownership, and location, their fundamental operations remain the same. The operations associated with a warehouse process (Fig. 10.1) include receiving, put away, storage, order picking, packing, loading, stock counting, value-adding services, and dispatch. The goal of most warehouses is to increase throughput rates and reduce the amount of stock held. Hence, cross docking where products are moved across the warehouse without actually going through the put-away process is handy to avoid the need to place the product into storage and any subsequent picking operation (Richards 2011).

Among the warehouse operations, order-picking refers to the operation of retrieving products or stock-keeping units (SKUs) from storage locations or buffer

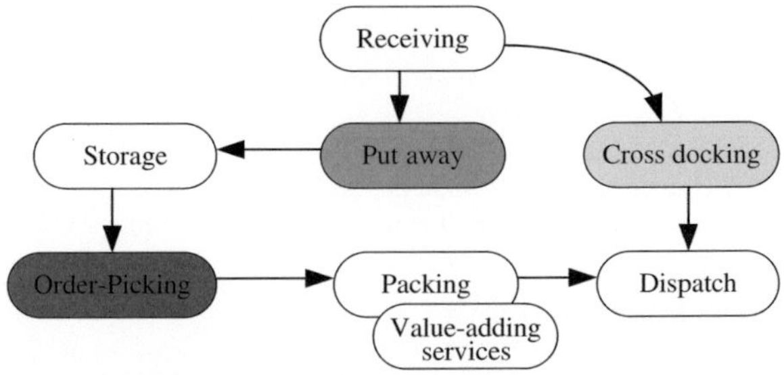

Fig. 10.1 Warehouse processes

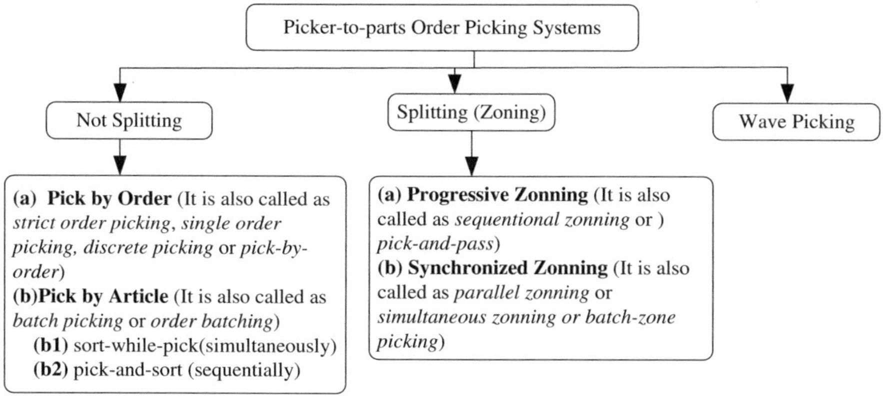

Fig. 10.2 Classification of picker-to-part order-picking systems

areas in response to fulfilling specific customer orders. It has long been appointed as the most labor-intensive and costly activity for almost every warehouse. The cost of order picking is estimated to be as much as 55 % of the total warehouse operating expense (DeKoster et al. 2007). Hence, order picking has been identified as the highest priority activity for productivity improvements (Thompkins et al. 2003).

DeKoster et al. (2007) classified order-picking systems according to whether humans or automated machines are utilized. The picker-to-part system, where the order picker walks or drives through the aisles to pick items, is the most common labor-intensive order-picking system (DeKoster et al. 2007). In Fig. 10.2, the picker-to-part order-picking systems are classified and maintained the terminological integrity for the names of the strategies based on the literature.

Picker-to-part systems can be classified as *low-level picking* and *high-level picking* based on the level of the storage racks or bins. It can be also classified based on the organization of the storage: not splitting and zoning–splitting the entire warehouse into multiple regions. If there is only one zone, then *discrete picking* or *bath picking* may be possible strategies to be employed.

In *discrete picking*, a picker is responsible for picking all items in a single order during a pick tour (Parikh and Meller 2008). This is also referred as *strict order picking, pick by order,* or *single order-picking.*

In *batch picking*, multiple customer orders are grouped together and picked in a given batch by an order picker. This is also called as *picking by article* or *order batching*. If the batch-picking strategy is selected, the picked orders must be sorted (Rouwenhorst et al. 2000). Two types of sorting policies exist: *pick-and-sort* (sequentially) and *sort-while-picking* (simultaneously). Under the *pick-and-sort* batch-picking policy, the pickers do not sort while picking the items to fulfill customer orders. Therefore, orders must be consolidated through a manual or automated sorting system. Consolidation refers to the grouping of items destined for the same customer. In contrast, under the sort-while-picking policy, the pickers performed these two processes simultaneously.

Zone picking requires that each picker be assigned to a specific region of the storage area and be responsible for picking items only in that region (Parikh and Meller 2008). Depending on the picking strategy, zoning may be further classified into two types, *progressive* (*pick-and-pass*) and *synchronized* (*simultaneous*) depending on whether the orders picked in a zone are passed to other zones for completion or picked in parallel (DeKoster et al. 2007). They are also called, respectively, *sequential* and *parallel zoning*. In *synchronized* zone picking, because all items corresponding to grouped orders are picked simultaneously from all zones, orders must be consolidated through a sorting system. Petersen (2000) calls synchronized zone picking as *bath-zone picking*.

Naturally, each strategy has advantages and disadvantages (Gu et al. 2007): Discrete order picking is simple to implement and not prone to mispicks, but it is labor-intensive; batch picking has a higher productivity due to less worker travel per item picked, but it can result in congestion and mispicks; zone picking has high productivity (if there is sufficient picking activity to keep workers busy), but it requires downstream sortation, which can be costly. Parikh and Meller (2008) summarized the advantages and disadvantages of batch and zone order-picking strategies based on the pick rate (items picked per unit time), the requirement of a sorting system, blocking, and the workload imbalance between pickers which directly affect the efficiency of the order-picking system in a warehouse.

In batch or zone picking, if the orders are required to be picked in a predefined time window (known as a wave), then it is referred to as *wave picking*. DeKoster et al. (2007) mentioned that it is usually (but not necessarily) combined with batch picking. Petersen (2000) defined it as a special case of synchronized zone picking. Wavelengths commonly range from 30 min to 2 h, and each picker picks continuously during the wave pausing only to unload the picking cart when it is full. Wave picking requires a picker to pick an item only once per wave, permitting even greater volume picking, whereas it requires more time and space for order consolidation because waves contain more orders than batches (Petersen 2000).

Within one warehouse, single or multiple order-picking methods can be employed. The selection of the appropriate strategy/strategies and the design of an effective order-picking operation are crucially important warehouse design and control problems. Over the last few decades, many papers (see in Rouwenhorst et al. 2000; DeKoster et al. 2007; Gu et al. 2007; Lodree et al. 2009) have been appeared studying order-picking processes. A few of them (Yoon and Sharp 1995; Dallari et al. 2009; Melacini et al. 2011) concentrated on designing an order-picking system theoretically. Application studies concentrated primarily on high-level and AS/RS systems (see examples in DeKoster et al. 2007).

In this study, a real-life, low-level, picker-to-part order-picking system employing humans with multiple picks per route in a 3PL firm is investigated. The 3PL firm has many customers. This study focuses on the orders of a specific customer who engaged in retail sales in the fashion industry. In the warehouse of the 3PL firm, the customer's fashion products are classified as put away and cross dock. Because the customer changed its policy and the percentages of the put-away and cross-dock products, the firm needs to design an order-picking system for this new arrangement. Hence, in this study, inspired by the methodology developed by Dallari et al. (2009), to choose the most appropriate order-picking system, a process flow consisting strategic decisions, operational implementation, and ES sections is developed. After utilizing the steps at the strategic decision level, in the operational level, five alternative order-picking strategies are utilized. For each alternative case, mathematical models adapted from Parikh and Meller's (2008) *analytical approach* for the order-batching problem are solved to maximize the total number of items fulfilled. In the ES section, the order-picking strategy having the minimum cost is selected after conducting accost analysis for alternative strategies.

The remainder of the paper is organized as follows: Sect. 10.2 presents a detailed literature review to highlight the contributions of this study. The problem definition is given in Sect. 10.3. Section 10.4 is devoted to the proposed design steps of the order-picking system. The steps utilized for the considered system are presented in Sect. 10.5. The final section presents conclusions and future directions.

10.2 Literature Review

The fashion and apparel industry is characterized by short product life cycles (Caro and Gallien 2015), high volatility and low predictability in demand (Christopher and Peck 1997), huge amount of product variety (Mehrjoo and Pasek 2014), and the requirement of the quick response to demand changes (Choi and Sethi 2010). In the fashion industry, the success of fast fashion retailers highly depends on the effective management of fashion retail supply chains. Choi (2014) defined fashion retail supply chain management as the planning and management of all activities involved in sourcing and procurement, conversion, and all logistics management activities in the fashion retail supply chain. Warehouse operations are critically important in logistics management activities of a fashion retail supply chain to maintain the effective management. Although the literature contains a huge amount of scientific studies on the decision problems for the design and control of warehouses, there are only limited studies (Caglioano et al. 2011; Choy et al. 2013; Meneghetti 2013; Hu et al. 2013) on warehouse operations in fashion companies. From a wide point of view, Rouwenhorst et al. (2000) presented a reference framework and a classification of warehouse design and control problems along three axes at a strategic, tactical, and operational level, respectively. They noted that the majority of papers are primarily analysis-oriented and do not pursue a synthesis of models and techniques as a basis for warehouse design.

The literature on order picking is also extensive. Lodree et al. (2009) reviewed the literature based on the physical and cognitive challenges associated with semiautomated order picking in warehouses. DeKoster et al. (2007) presented a literature overview on typical decision problems in the design and control of manual order-picking processes, which concerns some or all of the following problems at the operational level: batching, routing and sequencing, and sorting. Gu et al. (2007) investigated the literature and classified the studies in mathematical programming point of view: The *batching problem* as the partitioning of orders for assignment among the time slots (pick waves) is essentially a bin packing problem (BPP) and as partitioning of orders among the pickers is a variation of the classical vehicle routing problem (VRP) subject to some performance criteria and constraints such as picker effort, imbalance among pickers, time slots, picker capacity, and order due dates. The selected order-picking strategy plays an important role to determine the performance criteria. For example, if zone picking is employed, the batch should balance pick effort across the ones to achieve high picker utilization while minimizing pick time so that the number of pickers required is minimized. *The routing and sequencing* in order picking is a warehouse-specific traveling salesman problem (TSP), where the objective is to minimize the total material-handling cost when the picking/storing location of an item is given. For *the sorting problem* in addition to the some early simulation studies (Bozer and Sharp 1985; Bozer et al. 1988), an optimal assignment method to minimize the sorting time based on a set-partitioning model (Meller 1997) was reported.

Although there is a significant amount of literature related to the decision problems in order-picking systems at the operational level, only a few studies are interested in the selection of the appropriate order-picking strategy. Furthermore, they do not combine their selection approach with a process flow, which constituted the design steps of the order-picking system.

Herein, the literature is reviewed in detail for the studies which interested in picking system design and/or picking strategy selection based on the utilized techniques and picking strategies they considered. These studies are classified in Table 10.1. The first three of them are order-picking system design studies by Yoon and Sharp (1995, 1996), Dallari et al. (2009) and Melacini et al. (2011). Yoon and Sharp (1995) proposed a cognitive design procedure to determine an initial system configuration for OPS. It is limited to iterative top–down design steps. Afterward, Yoon and Sharp (1996) presented an original structured procedure for OPS design, based on iterative top–down decomposition and bottom–up modification. Dallari et al. (2009) developed a methodology to support warehouse designers in choosing the most suitable OPS. By developing a new OPS classification, they surveyed on over 68 distribution centers that had recently built in Italy. The results of the critical analysis allowed them to develop a design methodology to choose the most suitable OPS. This methodology has been integrated in the structured procedure for OPS design developed by Yoon and Sharp (1996). Melacini et al. (2011) developed a framework for (progressive zone picking) the pick-and-pass order-picking system design.

The rest of the studies mentioned in Table 10.1 are in a chronological order, as follows: Lin and Lu (1999), proposed a two-phased procedure to determine order-picking strategies. An analytic method was first employed to classify all orders into five categories based on the order quantities (EQ) and the number of order item (EN) in each order. A simulation software (Pro-Model) was utilized to generate the appropriate picking strategies that correspond to each type of order classified. There are two strategies used (Lin and Lu 1999): (1) single order-picking (SOP) and (2) batching and zoning (BZ). Through the SOP strategy, a picker travels

Table 10.1 Classification of the order-picking studies in the literature

Problem classification	Reference	Solution methodology	Picking strategies
Order-picking system design	Yoon and Sharp (1995, 1996)	Theoretical study	–
	Dallari et al. (2009)	Theoretical study (Case study without a integration of analytical methods)	–
	Melacini et al. (2011)	Theoretical study (Example problem)	Only for the (progressive zone picking) pick-and-pass order-picking system design
Picking strategy selection by performance evaluation based on simulation	Lin and Lu (1999)	Simulation (Example problem)	1. Single order picking 2. Bathing and zoning
	Petersen (2000)	Simulation and ANOVA (order-picking problem for Mail Order Companies)	1. Strict order 2. Batch 3. Sequential zone 4. Batch zone, and 5. Wave
	Petersen and Aase (2004)	Simulation Study (Example problem)	1. Strict order (S) 2. Batching (F)
	Yu and DeKoster (2009)	Queing network theory based model and simulation (Application)	1. Order batching and 2. Picking area zoning
Picking strategy selection by mathematical programming based analytical approach	Parikh ve Meller (2008)	Mathematical Programming based Analytical Approach (Example problem)	1. Batch picking (bulk) and 2. Zone picking
Picking strategy selection by performance evaluation based on simulation and DEA models	Chen et al. (2010)	Simulation and DEA (examples)	1. Single order picking and 2. Batching
Picking strategy selection by metaheuristics	Shen et al. (2001)	A GA-based heuristic approach (Experimental study)	1. Sequential zoning and 2. Simultaneous zoning

through the picking area with one order. Through the BZ strategy, a picker first combines some orders into one batch and then splits this batch into suborders for different picking zones. As a conclusion, there are new directions for the extension of the number of picking strategies employed. For instance, the strategies of relay with zoning and SOP with zoning are potential candidates for improving the picking performance. Another issue for enhancing picking performance is related to the number of pickers.

Petersen (2000) explored with the problem of order picking in mail-order companies. Five order-picking policies, strict order, batch, sequential zone, batch zone, and wave, were evaluated using labor requirements, processing time, and customer service as performance measures. A simulation model was developed to investigate these picking policies in a mail-order environment. Prior research had focused on the study of individual picking policies. This study extends the prior research by evaluating multiple picking policies under varying operating conditions. The results of the study seem to indicate that (1) wave picking and batch picking perform well across the range of operating conditions considered in this study and (2) sequential zone and batch zone picking do not perform well, especially when the order volume increases.

Petersen and Aaase (2004) examined the effect of three process decisions (picking, storage, and routing) on order picker travel, which is a major cost component of order fulfillment. They used a simulation model and considered different order-picking, storage, and routing strategies and performed sensitivity analyses to examine the effect of order size, warehouse shape, the location of pick-up/drop-off point, and demand distribution on performance. The results showed that the batching of orders yields the greatest savings, particularly when smaller order sizes are common. The results also showed that the use of either a class-based or volume-based storage policy provides nearly the same level of savings as batching, while being less sensitive to the average order size.

Parikh and Meller (2008) focused on the problem of selecting between a batch-picking (bulk) and a zone-picking strategy. For this problem, they proposed a cost model to estimate the cost of each type of picking strategy. In their cost model, they considered the effects of pick rate, picker-blocking, workload imbalance, and sorting system requirements. Through an example problem, they showed how system throughput, order sizes, item distribution in orders, and wavelength affect the picking strategy selection decision.

Yu and DeKoster (2009) proposed an approximation model based on queuing network theory to analyze the impact of order-batching and picking-area zoning on the mean order throughput time in a pick-and-pass order-picking system. It is a real application and simulation study. Order-batching and zoning of the picking area are two important factors that will influence order-picking efficiency. This paper uses a G/G/m queuing network approximation model to analyze the impact of batching and zoning on order-picking system performance with online order arrivals.

Chen et al. (2010) evaluated the performance of order-picking systems with different combinations of storage and order-picking policies with a DEA-based approach.

Shen et al. (2001) presented a heuristic method based on a genetic algorithm for selection between sequential zoning and simultaneous zoning for the picker-to-part system. Five different order/stock-keeping unit (SKU) matrices with different densities "d" and quantities "q" following the uniform distribution were created in order to test the suitability of sequential zoning and simultaneous zoning. As the result of the experimental study, the optimal order-picking time for each zoning policy was obtained.

The review of the literature shows that a few of the studies (Yoon and Sharp 1995; Dallari et al. 2009; Melacini et al. 2011) concentrated on designing an order-picking system theoretically. At the same time when the strategy selection studies are investigated, most of them, other than the theoretical studies, are simulation studies. To the best of our knowledge, there is no a real-life application study that proposes a process flow for the design of order-picking systems with detailed implementation steps and utilizes these steps simultaneously. Furthermore, the current study incorporates an analytical approach and mathematical models into the proposed process flow to determine the optimal order-picking strategy with a cost-centric evaluation. These new features highlight the originality of this paper.

10.3 System Analysis and Problem Description

In this section, some more details of the case study firm and the problem are reported. The 3PL firm, which was established in 1990, has many customers from different industrial sectors such as automotive, fashion and textile, and health care. The customer, whose orders are subject to this study, is primarily engaged in retail sales in the fashion industry. The customer is one of the well-established fashion apparel companies in Turkey. Customer satisfaction is the primary motivation of the company. Because this fashion company's own operations are highly specialized on fashion, it prefers to work with a logistics service provider to help with receiving the orders from the sale management and sending the ordered products to the finished product store in the agreed times and quantities. In the warehouse of the logistic firm, the customer's folded fashion and apparel products are classified as put away and cross dock. The customer makes orders by transferring the daily order requirements for products to the 3PL firm's warehouse management system (WMS). After receiving orders, the grouping, allocation, sequencing and routing, picking, and sorting processes are executed.

By 2014, the total volume of the overall products had increased and the customer's storage policy was changed. The percentages of the put-away and cross-dock products coming from the customer were reversed to 30 and 70 %, respectively. In the previous state, the order-picking operation was carried out on a group basis. In other words, each picker picked the entire basket, which contained at least one ordered product. After picking, the ordered products were sorted based on the store basis using the flow rack system, and the remaining products in the baskets were addressed again in the stock area. Due to the new conditions imposed on the order quantity, the products' quantity in the orders, the logistic firm needed to design an order-picking system. To increase the pick rate of pickers, the pick-and-sort batch picking and the synchronized zone-picking strategies were chosen for the alternatives to design an order-picking system. Furthermore, the current rack system was appropriate to utilize for both strategies. Additionally, the rest of the order-picking equipments will be the same for the different levels of each strategy.

Throughout this study, the currency of the cost data is Turkish Liras (TRY). Based on Turkey's Central Bank's currency rates as of January 1, 2014, one USD is worth 2.19 TRY. Each picker needs one handheld barcode scanner for the order-picking process.

The customer shares the annual forecasts of the order-based average daily folded fashion product quantities with the 3PL firm. Order-picking operations are carried out in working hours 08:00 a.m. to 6:00 p.m. per day. The lunch break is 1 h, and there are two shifts of 4 h and 30 min per day. In other words, there are two time windows/waves per day. The daily orders received are required to be fulfilled up until 11.00 p.m. of the associated day. For incomplete orders, overtime work is required.

10.4 Solution Methodology

10.4.1 The Proposed Process Flow

In this study, enhanced process flow is proposed for the order-picking system design. This process flow can be adopted for any order-picking design and selection problem with slight modifications based on the considered system. The flow consists primarily of three levels: the strategic decision level, the operational level, and the ES level. Strategic decisions have a long-term impact; they remain same as long as the customer's demands; and the type of warehousing system has no radical alterations. At the operational level, allocation, sequencing, routing, and sorting decisions are operated after the identification of alternative strategies for order picking, and the orders are grouped based on mathematical programming. The ES level constitutes cost-centric analytical approach to select the appropriate order-picking strategy.

Steps of the Proposed Process Flow:

Strategic Decision Level

1. Analyze the customer's products.
2. Analyze the customer's orders.
3. Need to change the existing layout or capacity? If the answer is yes, then go to step 4. If the answer is no, then go to step 5.
4. Calculate the capacity and plan the layout appropriately. Then go to step 5.
5. Discover the layout properties to use for the remaining steps. Analyze the existing routing policy.
6. Need to change the existing routing policy? If the answer is yes, then go to step 7. If the answer is no, then go to step 8.
7. Solve the routing problem by using the most appropriate method for the system. Then go to step 6.
8. Analyze the existing storage policy.
9. Need to change the existing storage policy? If the answer is yes, then go to step 10. If the answer is no, then go to step 11.

10. Determine the most appropriate storage policy for the system.
11. Identify the grouping strategy to meet the needs.
12. Identify the allocation strategy to meet the needs.
13. Analyze the sorting system.

Operational Level

14. Identify the appropriate order-picking strategy alternatives.
15. Perform work studies to identify the number of items to be picked in 1 man-hour for batch picking and zone picking.
16. Group the orders according to grouping strategies for each alternatives.
17. Make the allocation according to allocation strategies.
18. Perform the sequencing and routing.
19. Perform the order picking for each alternatives.
20. Perform the sorting.

Evaluation and Selection Level

21. Calculate the total cost for each alternatives.
22. Select the order-picking strategy which has the minimum cost and apply the following orders.

10.4.2 Utilized Mathematical Models

To calculate the cost of different systems corresponding to each order-picking strategy, Parikh and Meller (2008) proposed a logical cost model that includes the expected cost of pickers, equipment, imbalance, sorting, and packers. In the current study, Parikh and Meller's model is revised by neglecting the sorting and packaging costs, because sorting and packaging operations are carried out exactly the same in both strategies in respective operational stages in the considered real system.

The revised logical model for the least expensive picking system is as follows:

$$\text{Min C} = \min\{C^{\text{b}}, C^{\text{z}}\} \tag{10.1}$$

subject to:

$$C^{\text{b}} = P^{\text{b}}C_{\text{p}} + C_{\text{e}}^{\text{b}} + U^{\text{b}}C_{\text{u}} \tag{10.2}$$

$$C^{\text{z}} = P^{\text{z}}C_{\text{p}} + C_{\text{e}}^{\text{z}} + U^{\text{z}}C_{\text{u}} \tag{10.3}$$

$$C^{\text{b}}, C^{\text{z}}, P^{\text{b}}, C_{\text{p}}, C_{\text{e}}^{\text{b}}, U^{\text{b}}, C_{\text{u}}, P^{\text{z}}, C_{\text{e}}^{\text{z}}, U^{\text{z}} \geq 0 \tag{10.4}$$

where C^b and C^z represent batch and zone-picking system costs, respectively. Notations in Eqs. (10.1)–(10.4) are defined as follows.

$\boldsymbol{P^b}$ The total number of pickers required in batch picking
$\boldsymbol{P^z}$ The total number of pickers required in zone picking
$\boldsymbol{C_p}$ The annualized loaded cost of a picker
$\boldsymbol{C_e^b}$ The annualized equipment cost in batch picking
$\boldsymbol{C_e^z}$ The annualized equipment cost in zone picking
$\boldsymbol{U^b}$ The total number of items in orders unfulfilled due to workload imbalance in a day in batch picking
$\boldsymbol{U^z}$ The total number of items in orders unfulfilled due to workload imbalance in a day in zone picking
$\boldsymbol{C_u}$ The annualized cost of workload imbalance associated with overtime of pickers.

In this logical model, the cost components associated with pickers and equipment are calculated straightforwardly. To calculate the cost of workload imbalance associated with the overtime of pickers, Parikh and Meller's (2008) order-batching models for batch-picking and zone-picking strategies are utilized to group the orders. The notations and models (Table 10.2) for each strategy are as follows:

Index set

$\boldsymbol{i}$ = index for orders	$i = 1,\ldots, D$
$\boldsymbol{j}$ = index for pickers	$j = 1,\ldots, P$
$\boldsymbol{l}$ = index for zones	$l = 1,\ldots, Z$
$\boldsymbol{k}$ = index for time windows	$k = 1,\ldots, W$

Parameters

$\boldsymbol{d_i}$ parameter for the number of items in order i
$\boldsymbol{d_{il}}$ parameter for the number of items in order i to be picked from zone l

Table 10.2 Order-batching models for batch picking and zone picking

Order-batching model for batch-picking strategy		Order-batching model for zone-picking strategy	
$\text{Max}\ z^b = \sum_{i=1}^{D}\sum_{J=1}^{P}\sum_{k=1}^{W} d_i x_{ijk}$	(10.5)	$\text{Max}\ z^z = \sum_{i=1}^{D}\sum_{l=1}^{Z}\sum_{k=1}^{W} d_{il} x_{ik}$	(10.11)
s.t.		s.t	
$\sum_{i=1}^{D} d_i x_{ijk} = b_{jk} \quad \forall j,k,$	(10.6)	$\sum_{i=1}^{D} d_{il} x_{ik} = b_{lk} \quad \forall l,k,$	(10.12)
$b_{jk} \leq M^b \quad \forall j,k,$	(10.7)	$b_{lk} \leq M^z \quad \forall l,k,$	(10.13)
$\sum_{J=1}^{P}\sum_{k=1}^{W} x_{ijk} < 1 \quad \forall i,$	(10.8)	$\sum_{k=1}^{W} x_{ik} \leq 1 \quad \forall i,$	(10.14)
$x_{ijk} \epsilon \{0,1\} \quad \forall i,j,k,$	(10.9)	$x_{ijk} \epsilon \{0,1\} \quad \forall i,j,k,$	(10.15)
$d_i, b_{jk}, M^b \geq 0 \quad \forall i,j,k,$	(10.10)	$d_{il}, b_{lk}, M^z \geq 0 \quad \forall i,l,k,$	(10.16)

M^b The maximum number of items that can be picked by a picker in a time window in a batch-picking system

M^z The maximum number of items that can be picked by a picker in a time window in zone-picking system.

Variables

x_{ijk} $\begin{cases} 1, & \text{if order } i \text{ is assigned to picker } j \text{ in a time window } k, \\ 0, & \text{otherwise} \end{cases}$

x_{ik} $\begin{cases} 1, & \text{if order } i \text{ is assigned to time window } k, \\ 0, & \text{otherwise} \end{cases}$

b_{jk} the current number of items assigned to each picker in a time window (capacity of a picker)

b_{lk} the current number of items assigned to each zone in a time window.

Each objective function ((10.5), (10.10)) is to maximize the number of the items fulfilled during the scheduled hours of operation. Constraint (10.6) determines the current number of items assigned to each picker in a time window. Constraint (10.7) guarantees that no more items are assigned to a picker beyond its capacity. Constraint (10.8) prevents the assignment of an order in any time window to more than one picker. Constraints (10.9), (10.10), (10.15), and (10.16) represent binary variables and positivity conditions.

The total number of items corresponding to the unfulfilled orders (U^b) placed in the logical cost model ((10.1)–(10.4)) is calculated as the difference between the total number of items to be picked and the optimal objective value (z^b) of the model ((10.5)–(10.10)) as follows:

$$U^b = \sum_{i=1}^{D} d_i - \sum_{i=1}^{D} \sum_{J=1}^{P} \sum_{k=1}^{W} d_i x_{ijk} \tag{10.17}$$

Constraint (10.12) specifies the number of items assigned to each zone in a time window. Constraint (10.13) prevents the number of items assigned to a zone exceeds the capacity of the respective zone. Constraint (10.14) ensures that no order is assigned to more than one time window.

The total number of items corresponding to the unfulfilled orders (U^z) placed in the logical cost model ((10.1)–(10.4)) is calculated as the difference between the total number of items to be picked and the optimal objective value (z^z) of model ((10.11)–(10.16)) as follows:

$$U^z = \sum_{i=1}^{D} \sum_{l=1}^{Z} d_{il} - \sum_{i=1}^{D} \sum_{l=1}^{Z} \sum_{k=1}^{W} d_{il} x_{ik} \tag{10.18}$$

10.5 Implementation

10.5.1 Strategical Decisions

- *Analyze the customer's products*:
  There are five categories of the folded fashion group: female, male, child, youth, and sporty. The stock percentages of each respective category in the customer's order quantity are 30 % (female), 25 % (male), 20 % (child), 15 % (youth), and 10 % (sporty).
- *Analyze the customer's orders*:
  The substitution of the products sold daily in stores is supplied from the storage inventory, which constitutes suppliers' put-away products. The product diversity of orders is at a high level, and the average number of products picked from an address in the storage area during the order-picking operation is four.
  The customer has 64 stores to which transfer is made for the folded fashion group, and these stores send their orders to the storage daily. The number of products up to 14,000 pieces should be provided from the warehouse within 24 h. All orders are gathered up until 11.00 p.m. After this hour, a WMS-based computer program is run to combine all orders, ensuring that the total amount does not exceed 14,000 pieces of product to obtain 64 pieces of store-based orders. Orders are determined according to the "first-in, first-out" rule. Orders that exceed 14,000 pieces are considered for the next day.
  The average numbers of products, which are ordered daily from each store, are given in the second column of Table 10.3. The customer does not make the category breakdown while sharing the prediction of the following orders within the year. In this study, it is assumed that the percentage of stock categories and the category percentage of orders are directly proportional. This assumption is also confirmed by the customer. The store-based and category-based daily average orders of pieces of products are calculated using the category percentages in the stock area. They are given in the last five columns of Table 10.3.
- *Need to change the existing layout or capacity? If the answer is no, then go to step 5.*
  The stock area considered in this study was previously organized based on the customer orders for the year 2014. It is 1536 square meters (m^2) in size and has 216,000 item capacity. The firm wants to utilize the same layout without a retrofit for economic reasons.
- *Discover the layout properties to use for the remaining steps.*
  The rectangular warehouse layout fits for a low-level, picker-to-part order-picking system employing humans with multiple picks per route. There are 18 parallel picking aisles running between the north and south walls and two storage blocks separated with a cross aisle. Each parallel picking aisle is two-sided with 15 pick locations on each side and is 1.4 m wide, which is enough to allow for two-way travel for two pickers with their pick carts (Fig. 10.3). The width of the cross aisle between two storage blocks and of the

Table 10.3 The store-based and category-based daily average order of pieces of product

Store code	Order based average daily folded product quantity	Female	Male	Child	Youth	Sporty
1	620	186	155	124	93	62
2	480	144	120	96	72	48
3	420	126	105	84	63	42
4	450	135	112	90	67	45
5	420	126	105	84	63	42
6	280	84	70	56	42	28
7	530	159	133	106	80	53
8	340	102	85	68	51	34
9	260	78	65	52	39	26
10	280	84	70	56	42	28
11	240	72	60	48	36	24
12	220	66	55	44	33	22
13	410	123	102	82	61	41
14	360	108	90	72	54	36
15	160	48	40	32	24	16
16	150	45	38	30	23	15
17	120	36	30	24	18	12
18	180	54	45	36	27	18
19	90	27	22	18	13	9
20	150	45	38	30	23	15
21	210	63	52	42	31	21
22	180	54	45	36	27	18
23	190	57	47	38	29	19
24	250	75	63	50	37	25
25	260	78	65	52	39	26
26	90	27	22	18	14	9
27	190	57	48	38	28	19
28	250	75	62	50	38	25
29	220	66	55	44	33	22
30	230	69	58	46	34	23
31	270	81	67	54	41	27
32	260	78	65	52	39	26
33	130	39	33	26	19	13
34	140	42	35	28	21	14
35	260	78	65	52	39	26
36	280	84	70	56	42	28
37	50	15	12	10	8	5
38	60	18	15	12	9	6
39	190	57	48	38	28	19

(continued)

Table 10.3 (continued)

Store code	Order based average daily folded product quantity	Female	Male	Child	Youth	Sporty
40	90	27	22	18	14	9
41	140	42	35	28	21	14
42	160	48	40	32	24	16
43	230	69	58	46	34	23
44	240	72	60	48	36	24
45	230	69	57	46	35	23
46	230	69	58	46	34	23
47	100	30	25	20	15	10
48	40	12	10	8	6	4
49	180	54	45	36	27	18
50	160	48	40	32	24	16
51	230	69	57	46	35	23
52	190	57	48	38	28	19
53	140	42	35	28	21	14
54	230	69	57	46	35	23
55	180	54	45	36	27	18
56	160	48	40	32	24	16
57	170	51	43	34	25	17
58	90	27	22	18	14	9
59	160	48	40	32	24	16
60	180	54	45	36	27	18
61	130	39	33	26	19	13
62	180	54	45	36	27	18
63	160	48	40	32	24	16
64	130	39	33	26	20	13
Total	14,000	4200	3500	2800	2100	1400

end cross aisles located parallel to the north and south walls of the warehouse is 1.8 m wide. The warehouse consists of (15 × 2 × 18=) 540 picking locations. Products are placed in baskets addressed to the locations. Each address has four baskets. Each basket has an RF code (barcode). The input and output (I/O) point is located in the lower right-hand corner in the layout.

- *Analyze the existing routing policy.*
  The locations in the vertical aisle are numbered from 1 to 15. The racks opposite each other in the same aisle have the same address numbers, but they are titled as "right" and "left." The address numbers of racks in an aisle are in the ascending order from the same direction. The utilized WMS system in the company ensures the progress of the picker's movement in an aisle by picking the items one by one from right and left racks toward the end of the aisle. After picking the ordered items in one aisle, the picker proceeds in the opposite

Fig. 10.3 Pick cart for folded fashion products

direction of the previous aisle to pick the ordered items in the new aisle. Thus, the picking process occurs by drawing an "S," which is the policy of S-shaped (traversal) routing.

- *Need to change the existing routing policy? If the answer is no, then go to step 7.*
  In practice, one of the most commonly applied routing policies is the S-shape (traversal) policy. DeKoster et al. (2007) summarized the commonly used methods for routing order pickers in single-block warehouses. Roodbergen and DeKoster (2001) modified these methods to be employed in multiple block warehouses. Hwang et al. (2004) compared the performances of three routing policies, and they indicated that the S-shape policy performs better for large order sizes.
  In the folded fashion product group examined in this study, in accordance with the high volume of orders, pickers are required to go into every aisle. Hence, the S-shape routing policy is determined to be the appropriate routing policy.
- *Analyze the existing storage policy.*
  In the current state, items are stored according to the category-based classification. Each category is represented with a different color rack where yellow_(child), red_(female), blue_(male), pink_(sporty), and green_(youth) racks are shown in Fig. 10.4. The distribution of 540 locations based on stock percentages of the respective categories is 165 (female), 135 (male), 105 (child), 81 (young), and 54 (sporty). Based on the stock percentage of each respective category and S-shape routing policy, the average numbers of items in one aisle to be picked daily are calculated by rule-of-thumb calculations and given in Table 10.4.

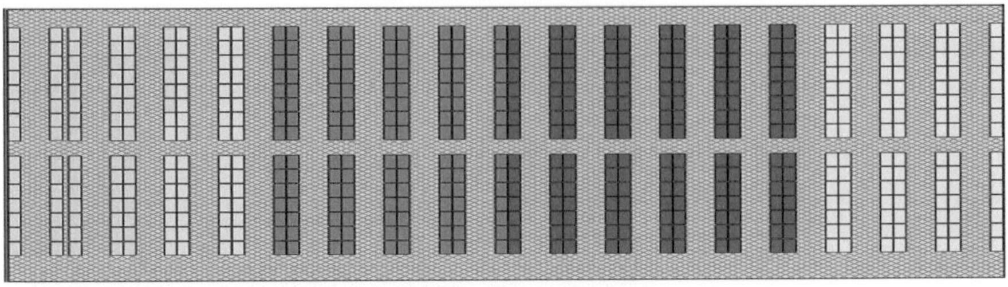

Fig. 10.4 The existing product category allocation in the storage layout

Table 10.4 Order-based average daily folded product quantities of each category

Aisle	Order-based average daily folded product quantity	Female	Male	Child	Youth	Sporty	Total workload in aisle
1	800	–	–	800	–	–	800
2	800	–	–	800	–	–	800
3	800	–	–	800	–	–	800
4	785	385	–	400	–	–	785
5	763	763	–	–	–	–	763
6	763	763	–	–	–	–	763
7	763	763	–	–	–	–	763
8	763	763	–	–	–	–	763
9	763	763	–	–	–	–	763
10	777	–	777	–	–	–	777
11	777	–	777	–	–	–	777
12	777	–	777	–	–	–	777
13	777	–	777	–	–	–	777
14	781	–	392	–	–	389	781
15	778	–	–	–	–	778	778
16	777	–	–	–	544	233	777
17	778	–	–	–	778	–	778
18	778	–	–	–	778	–	777
Total	14,000	4200	3500	2800	2100	1400	

- *Need to change the existing storage policy? If the answer is yes, then go to step 9*

  The existing storage policy should be changed because it causes workload imbalance. The layout plan is presented in Fig. 10.4. Aisles 4, 14, and 16 in Table 10.4 have two different categories, whereas there is a single category for the rest of the aisles. By looking at the order-based average daily folded product quantities of each category in Table 10.4, the highest workload aisles are 1, 2, 3

with 800 items and the lowest workload aisle is 9 with 763 items. The difference between highest and lowest number of picked items in an aisle is 37.
According to the time study conducted on the batch order-picking strategy, there was a 15.3 min difference between the processing times of the picking operations in the aisles. Another time study was performed on the zone order-picking strategy when the area size was restricted to 1 aisle, and 2 aisles and 3 aisles wide. A 13 min difference, 13.6 min difference, and 14.2 min difference, respectively, found between the processing times of picking operations in aisles for zoning. The workload imbalance between the aisles is critical for both order-picking strategies.

- *Determine the appropriate storage policy for the system.*
The literature traditionally deals with the storage allocation problem and the storage assignment problem separately (Accorsi et al. 2012). Storage allocation determines the right inventory level per each SKU, and storage assignment addresses the assignment of SKU to the most convenient locations. The literature presents several storage assignment strategies which are classified mainly as *randomized, dedicated,* and *class-based.* In the *randomized* storage, assignment policy, products/stock-keeping units are arbitrarily assigned to the first available location in the order-picking system. The *dedicated* storage strategy assigns an item based on known rules which are ranked index-based, correlated storage assignment-based. In the *class-based* storage strategy, the products are divided into a number of categories according to their demand, and each category is associated with a set of zones where the products are stored according to a random storage policy. The class-based storage strategy reduces to the dedicated storage policy if the number of categories is equal to the number of products and to the randomized storage strategy if there is a single category (Ghiani et al. 2013). Roodbergen (2012) indicated four variations of class-based storage: across-aisle storage, within-aisle storage, nearest-subaisle storage, and nearest-location storage. The ranked index-based rules use different criteria to assign a product to a storage location (Gu et al. 2007). These criteria are popularity, maximum inventory, order closing, turn, and cube per order index. The correlated storage assignment policy, which is also known as family grouping, locates the SKUs with a high degree of correlation close to each other.
In this study apart from the above-mentioned strategies, to obtain an improvement in terms of workload imbalance between aisles, more than one category into the same aisle is addressed based on the order percentages of the categories. To generate this addressing arrangement, the number of racks required for each category is calculated and the order percentages is accepted as the weight of the related rack. That is, taking 30 (female), 25 (male), 20 (child), 15 (youth), and 10 (sporty) as weights for each respective category, the racks belonging to the categories have been placed in the storage yard by trying to balance the total weight of the aisle. The storage layout of the category breakdown that will emerge as a result of the planning is shown in Fig. 10.5. In which, again,

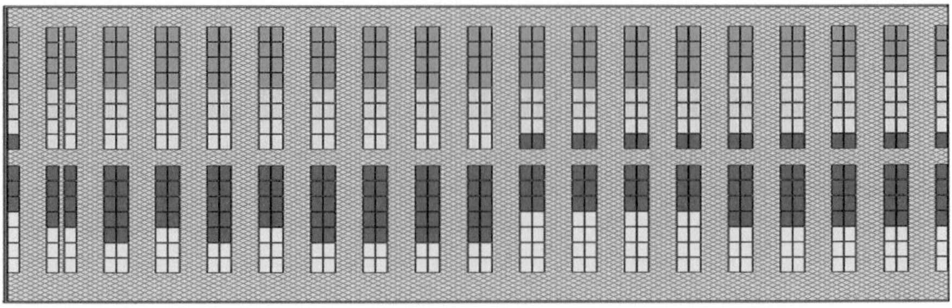

Fig. 10.5 The new product category allocation in the storage layout (Color figure online)

Table 10.5 Order-based average daily folded product quantities of each category after the development

Aisle	Order-based average daily folded product quantity	Female	Male	Child	Youth	Sporty
1	778	254	156	160	156	52
2	778	254	156	160	156	52
3	777	254	155	160	156	52
4	778	254	156	160	156	52
5	777	229	181	187	128	52
6	778	204	205	214	104	51
7	778	204	207	214	102	51
8	778	204	205	214	103	52
9	778	229	207	160	104	78
10	777	255	208	106	104	104
11	777	255	208	106	104	104
12	778	255	208	107	104	104
13	778	229	208	133	104	104
14	778	229	208	133	104	104
15	778	229	208	133	104	104
16	778	229	208	133	104	104
17	778	229	208	133	104	104
18	778	204	208	187	103	76

yellow, red, blue, pink, and green racks dedicated to the child, female, male, sporty, and youth categories, respectively.

The average piece of product that is needed to be picked per aisle and its dispersion to each category are shown in Table 10.5. As an example of the rule of thumb, the number of items from the female category to be picked in the first aisle is calculated as follows: There are 165 red areas in total, 10 of which placed in the first aisle from the left in Fig. 10.5. The total number of female items to be picked is 4200 in a day which is given in Table 10.3. Hence, in the

first aisle from the left in the layout (Fig. 10.5) (4200 × 10/165) = 254 items are stored.
It is evident that addressing all categories in the same aisle based on the order percentage of each category improves the workload balance between aisles. In aisles 3, 10, and 11, the order-based average daily folded product quantities, which correspond the total workloads in respective aisles, are equal to 777, and in the rest of the aisles, the total workload is 778. Hence, the difference between the highest workload and the lowest workload is 1, based on the aisles. The difference of this 1 item causes a 0.4 min difference between the processing times of picking operations between the aisles for both considered order-picking strategies.

- *Identify the grouping strategy to meet the needs.*
  Based on the new storage policy (Fig. 10.5), the proper designation of products in the warehouse prohibits workload imbalance between aisles. After this stage, ensuring workload balance during the grouping of orders comes next. To group the orders, order-batching models (Table 10.2) are utilized under the assumptions of the considered order-picking strategies at the operational level.
- *Identify the allocation strategy to meet the needs.*
  The allocation algorithm followed in the company is managed by the WMS system: If a product is found at more than one address, then the numbers of units in baskets located in related addresses are compared by the system and the allocation process is performed by placing the products in the basket with the lowest number of units.
- *Analyze the sorting system.*
  Within the warehouse, the sorting process is accomplished using flow racks. Each part of the flow rack represents a storehouse. The products are sent out to the compartments on one side of the flow racks, and the products are packaged at the packing tables on the other side of them.

10.5.2 Operational Application

- *Identify the appropriate order-picking strategy alternatives.*
  Because of the customer's order size, the average daily orders and the number of stores are high, and the pick-and-sort batch-picking strategy with two different levels for the picker number and the synchronized zone-picking strategy with three different zone sizes are utilized as five alternative order-picking strategies based on the considered problem. For each alternative case, the mathematical model of the order-batching problem is solved to maximize the total number of items fulfilled.
- *Perform work studies to identify the number of items to be picked in 1 man-hour for batch picking and zone picking.*
  To determine the number of items to be picked in 1 man-hour, time studies (stop-watch measurement analyses) are performed for batch picking and zone

picking strategy alternatives. One hundred forty-five items/man-hour for batch picking, 171 items/man-hour for zone picking with one aisle, 163 items/man-hour for zone picking with two aisles, and 156 items/man-hour for zone picking with three aisles are obtained.

- *Group the orders according to grouping strategies for each alternatives.*
  Mathematical model results for batch-picking strategy
  The time windows defined for order picking consist of two shifts with 4.5-h periods ($k = 2$). In this case, an employee works for 9 net hours per day doing order picking. Because there are 14,000 pieces ordered daily that should be picked, when the batch order-picking strategy is applied, theoretically (14,000 items/145 items per man-hour) = 96.5 man-hours is needed which requires ($96.5/9 \cong 10.7$) 11 pickers ($P = 11$) who work daily. The maximum number of items a picker can pick in a time window is (4.5 h * 145 items/man-hour = 652.5) 653 ($\boldsymbol{M}^{\mathbf{b}} = 653$). Because the customer has 64 stores that send their orders to the storage daily, the i index for orders is defined from 1 to 64 ($D = 64$). The data for the $\boldsymbol{d}_i$ parameter which is the number of items in order i are mentioned in Table 10.3. The order-batching mathematical model ((10.5)–(10.10)) for the batch-picking strategy is constructed based on the mentioned data. The optimal solution of the mathematical model is obtained with the GAMS computer package. Based on the optimal order-grouping solution summarized in Table 10.6, orders that belong to the store 1 and store 7 are not assigned to any of the pickers in any time window, which is indicated with green color rows in Table 10.6. As a result, not 14,000 but only 12,850 pieces of product in total can be delivered.
  By rearranging the optimal order-grouping solution given in Table 10.6, the total number of products and the capacity utilization rates that the pickers will pick in each time window are calculated and given in Table 10.7. As an example for the calculations in this rearrangement, picker number 1 in the first time window should pick (220 + 180 + 170=) 570 items, which is indicated with blue rows in Table 10.6, and placed in the first row of Table 10.7 with 11 pickers. Because the maximum number of items a picker can pick in a time window is 653, the capacity utilization for the case with picker number 1 in the first shift is calculated as (570/653) = 87.3 % which is indicated in Table 10.7. The average of capacity utilization rates of the pickers is calculated as 89.4 %, and the minimum capacity utilization rate of the pickers is 68.9 % in Table 10.7. Because two orders that are indicated with green rows in Table 10.6 ($i = 1$ and $i = 7$) are not assigned, the necessity of overtime work arises for 1150 (=620 + 530) pieces of product. Another alternative to deal with the problem is to increase the number of pickers. The same mathematical model is solved for 12 pickers when all other data remain the same. The results indicate that the order with 150 items coming from the store number 20 is not assigned to any of the pickers in any time window, and 13,850 pieces of product in total can be delivered. Because of the space limitation, the optimal order-grouping solution summary is not placed herein. However, the total number of products and the

Table 10.6 Optimum grouping according to batch-picking strategy with 11 pickers

i	Product quantity	Picker *j*	Time windows *k*	*i*	Product quantity	Picker *j*	Time windows *k*
1	**620**	–	–	33	130	2	2
2	480	6	1	34	140	6	2
3	420	3	2	35	260	5	1
4	450	5	2	36	280	9	2
5	420	6	2	37	50	11	2
6	280	3	1	38	60	11	2
7	**530**	–	–	39	190	7	1
8	340	9	2	40	90	2	1
9	260	8	2	41	140	10	2
10	280	7	2	42	160	6	1
11	240	2	1	43	230	3	2
12	220	1	2	44	240	11	2
13	410	8	1	45	230	11	1
14	360	9	1	46	230	2	1
15	160	11	1	47	100	11	2
16	150	4	1	48	40	2	1
17	120	4	1	49	180	10	2
18	180	5	2	50	160	2	2
19	90	3	1	51	230	2	2
20	150	1	2	52	190	9	1
21	210	4	2	53	140	5	1
22	180	4	2	54	230	11	1
23	190	4	2	55	**180**	**1**	**1**
24	250	3	1	56	160	10	1
25	260	10	1	57	**170**	**1**	**1**
26	90	10	1	58	90	4	1
27	190	7	1	59	160	8	2
28	250	7	2	60	180	8	1
29	**220**	**1**	**1**	61	130	5	1
30	230	1	2	62	180	8	2
31	270	7	1	63	160	4	1
32	260	10	2	64	130	10	1

capacity utilization rates that the pickers will pick in each time window are calculated and given in Table 10.7. The average of capacity utilization rates of the pickers for the 12-picker case is calculated as 88.4 %, and the minimum capacity utilization rate of the pickers is 64.3 % in Table 10.7.

Table 10.7 Pickers' workload and capacity utilizations for batch-picking strategy with 11 and 12 pickers

11 pickers				12 pickers			
Picker	Time window	Assigned total product quantity	Capacity utilization (%)	Picker	Time window	Assigned total product quantity	Capacity utilization (%)
1	1	570	87.3	1	1	630	96.5
1	2	600	91.9	1	2	650	99.5
2	1	600	91.9	2	1	650	99.5
2	2	520	79.6	2	2	650	99.5
3	1	620	94.9	3	1	590	90.4
3	2	650	99.5	3	2	610	93.4
4	1	520	79.6	4	1	610	93.4
4	2	580	88.8	4	2	610	93.4
5	1	530	81.2	5	1	580	88.8
5	2	630	96.5	5	2	620	94.9
6	1	640	98.0	6	1	630	96.5
6	2	560	85.8	6	2	490	75.0
7	1	650	99.5	7	1	640	98.0
7	2	530	81.2	7	2	610	93.4
8	1	590	90.4	8	1	650	99.5
8	2	600	91.9	8	2	620	94.9
9	1	550	84.2	9	1	530	81.2
9	2	620	94.9	9	2	500	76.6
10	1	640	98.0	10	1	640	98.0
10	2	580	88.8	10	2	650	99.5
11	1	620	94.9	11	1	250	38.3
11	2	450	68.9	11	2	380	58.2
				12	1	420	64.3
				12	2	640	98.0
Average			89.4				88.4

10.5.2.1 Mathematical Model Results for Zone-Picking Strategy

The time windows defined for order picking are comprised of two shifts with 4.5-h periods (k = 2). The maximum number of items that one picker can pick in a time window is (4.5 h * 171 items/man-hour ≅770) 770 ($\boldsymbol{M}^{\boldsymbol{z}}$ = 770) in case (a) where the area size is 1 aisle wide; in case (b) where the area size is 2 aisles wide, it is (4.5 h * 163 items/man-hour ≅734) 734 ($\boldsymbol{M}^{\boldsymbol{z}}$ = 734); and in case (c) where the area size is 3 aisles wide, it is (4.5 h * 156 items/man-hour) = 702 ($\boldsymbol{M}^{\boldsymbol{z}}$ = 702). Because the customer has 64 stores that send their orders to the storage daily, the i index for orders is defined from 1 to 64 (D = 64). The data for the $\boldsymbol{d}_{il}$ parameter, which is the number of items in order i to be picked from zone l (Z = 18), are given in Table 10.8. The order-batching mathematical model ((10.11)–(10.16)) for the

Table 10.8 Average piece of product related to the orders when the zone-picking strategy with 1 aisle

Order		Zone $l = 1, 2, \ldots, 18$																	
i	Daily order quantity	1	2	3	4	5	6	7	8	9	10	11	12	13	14	15	16	17	18
1	620	34	34	34	34	35	35	35	35	35	35	35	35	34	34	34	34	34	34
2	480	27	27	27	27	27	27	27	27	27	27	27	27	26	26	26	26	26	26
3	420	23	23	23	23	23	23	23	23	23	23	23	23	24	24	24	24	24	24
4	450	25	25	25	25	25	25	25	25	25	25	25	25	25	25	25	25	25	25
5	420	23	23	23	23	23	23	23	23	23	23	23	23	24	24	24	24	24	24
6	280	15	15	16	16	16	16	16	16	16	16	16	16	15	15	15	15	15	15
7	530	29	29	29	29	30	30	30	30	30	30	30	30	29	29	29	29	29	29
8	340	19	19	19	19	19	19	18	18	19	19	19	19	19	19	19	19	19	19
9	260	15	15	15	14	14	15	14	14	14	14	14	14	14	14	15	15	15	15
10	280	15	15	16	16	16	16	16	16	16	16	16	16	15	15	15	15	15	15
11	240	13	13	13	13	12	12	12	12	12	12	13	13	15	15	15	15	15	15
12	220	12	12	12	12	12	12	12	12	12	12	12	12	12	12	13	13	13	13
13	410	23	23	23	23	23	23	23	23	23	23	22	22	23	23	23	23	22	22
14	360	20	20	20	20	20	20	20	20	20	20	20	20	20	20	20	20	20	20
15	160	9	9	9	9	9	9	9	9	9	9	9	9	8	8	9	9	9	9
16	150	8	8	8	8	8	8	8	8	8	8	9	9	9	9	8	8	9	9
17	120	7	7	7	7	7	7	7	7	7	7	7	7	6	6	6	6	6	6
18	180	10	10	10	10	10	10	10	10	10	10	10	10	10	10	10	10	10	10
19	90	5	5	5	5	5	5	5	5	5	5	5	5	5	5	5	5	5	5
20	150	8	8	8	8	8	8	8	8	8	8	9	9	9	9	8	8	9	9
21	210	12	12	12	12	12	12	12	12	12	12	11	11	11	11	12	12	11	11
22	180	10	10	10	10	10	10	10	10	10	10	10	10	10	10	10	10	10	10
23	190	11	11	11	11	11	11	11	11	11	11	10	10	10	10	10	10	10	10

(continued)

Table 10.8 (continued)

Order		Zone $l = 1, 2, \ldots, 18$																	
i	Daily order quantity	1	2	3	4	5	6	7	8	9	10	11	12	13	14	15	16	17	18
24	250	14	14	13	14	13	13	13	13	14	14	14	14	15	14	14	14	15	15
25	260	15	15	15	15	15	15	14	14	14	14	15	15	14	14	14	14	14	14
26	90	5	5	5	5	5	5	5	5	5	5	5	5	5	5	5	5	5	5
27	190	11	11	11	11	11	11	11	11	11	11	10	10	10	10	10	10	10	10
28	250	13	13	13	13	14	14	14	14	14	14	14	14	14	14	14	14	15	15
29	220	12	12	12	12	12	12	13	13	13	13	12	12	12	12	12	12	12	12
30	230	13	13	13	13	13	13	13	13	13	13	13	13	13	13	12	12	12	12
31	270	15	15	15	15	15	15	15	15	15	15	15	15	15	15	15	15	15	15
32	260	15	15	15	15	15	15	14	14	14	14	15	15	14	14	14	14	14	14
33	130	7	7	7	7	7	7	7	7	7	7	7	7	7	7	8	8	8	8
34	140	7	7	7	7	7	7	8	8	7	7	8	8	8	8	9	9	9	9
35	260	15	15	14	14	15	15	14	14	14	14	15	15	15	15	14	14	14	14
36	280	15	15	16	16	16	16	16	16	16	16	16	16	15	15	15	15	15	15
37	50	3	3	3	3	3	3	3	3	3	3	3	3	3	3	2	2	2	2
38	60	3	3	3	3	3	3	3	3	3	3	3	3	4	4	4	4	4	4
39	190	10	11	10	10	11	11	11	11	11	11	10	10	11	12	10	10	10	10
40	90	5	5	5	5	5	5	5	5	5	5	5	5	5	5	5	5	5	5
41	140	7	7	7	7	8	8	8	8	7	7	7	7	8	8	9	9	9	9
42	160	9	9	9	9	8	8	9	9	9	9	9	9	9	9	9	9	9	9
43	230	13	13	13	13	13	13	13	13	13	13	13	13	13	13	12	12	12	12
44	240	13	13	13	13	13	13	13	13	13	13	13	13	13	13	14	14	15	15
45	230	13	13	13	13	13	13	13	13	13	12	12	13	13	13	13	13	12	12
46	230	13	13	13	13	13	13	13	13	13	13	13	13	13	13	12	12	12	12

(continued)

Table 10.8 (continued)

Order		Zone $l = 1, 2, \ldots, 18$																	
i	Daily order quantity	1	2	3	4	5	6	7	8	9	10	11	12	13	14	15	16	17	18
47	100	6	6	6	6	6	6	6	6	6	6	5	5	5	5	5	5	5	5
48	40	3	3	3	3	2	2	2	2	2	2	2	2	2	2	2	2	2	2
49	180	10	10	10	10	9	10	10	10	10	10	10	10	10	10	10	11	10	10
50	160	9	9	9	9	9	9	8	8	9	9	9	9	9	9	9	9	9	9
51	230	13	13	13	13	13	13	13	13	13	13	13	13	13	13	12	12	12	12
52	190	11	11	11	11	11	11	11	11	11	11	10	10	10	10	10	10	10	10
53	140	8	8	7	7	7	7	8	8	7	7	8	8	8	8	9	9	8	8
54	230	13	12	13	13	13	13	13	13	13	13	13	13	13	13	13	12	12	12
55	180	10	10	10	10	10	10	10	10	10	10	10	10	10	10	10	10	10	10
56	160	9	9	9	9	8	8	8	8	9	9	9	9	10	10	9	9	9	9
57	170	10	10	10	10	10	10	10	10	9	9	9	9	9	9	9	9	9	9
58	90	5	5	5	5	5	5	5	5	5	5	5	5	5	5	5	5	5	5
59	160	9	9	9	9	8	8	9	9	9	9	9	9	9	9	9	9	9	9
60	180	10	10	10	10	10	10	10	10	10	10	10	10	10	10	10	10	10	10
61	130	7	7	7	7	7	7	7	7	7	7	7	7	7	7	8	8	8	8
62	180	10	10	10	10	10	10	10	10	10	10	10	10	10	10	10	10	10	10
63	160	9	9	8	9	9	8	9	9	9	9	9	9	9	9	9	9	9	9
64	130	7	7	7	7	7	7	7	7	7	7	7	7	7	7	8	8	8	8
Total	14,000	778	778	777	778	777	778	778	778	778	777	777	778	778	778	778	778	778	778

Table 10.9 Optimum grouping based on the zone-picking strategy with 18 zones

i	Product quantity	Time windows	*i*	Product quantity	Time windows
1	620	1	33	130	2
2	480	2	34	140	2
3	420	2	35	260	1
4	450	2	36	280	1
5	420	2	37	50	1
6	280	1	38	60	2
7	530	1	39	190	1
8	340	1	40	90	1
9	260	1	41	140	1
10	280	1	42	160	1
11	240	1	43	230	1
12	220	1	44	240	1
13	410	2	45	230	1
14	360	1	46	230	1
15	160	1	47	100	2
16	150	2	48	40	2
17	120	2	49	180	1
18	180	1	50	160	1
19	90	2	51	230	1
20	150	2	52	190	1
21	210	1	53	140	2
22	180	1	54	230	1
23	190	1	55	180	1
24	250	1	56	160	1
25	260	1	57	170	1
26	90	1	58	90	1
27	190	1	59	160	1
28	250	1	60	180	1
29	220	1	61	130	1
30	230	1	62	180	1
31	270	1	63	160	1
32	260	1	64	130	1

zone-picking strategy is constructed based on the mentioned data. The optimal solution of the mathematical model is obtained by the GAMS computer package. Based on the optimal order-grouping solution summarized in Table 10.9, there is 1 picker in every area. In other words, there are 18 pickers in total. Orders are separated into two groups, which are those assigned to the first time window and those assigned to the second time window. All orders are assigned, and there is no need for overtime work.

Based on the optimal order-grouping solution (Table 10.9), the capacity utilization rates that the pickers will pick in any time window are calculated and given in Table 10.10.

For case (a), the average of capacity utilization rates of the pickers is 50.5 % and the minimum capacity utilization rate of the pickers is 23.5 %. In this case, although all orders can be picked within the day, the pickers spend nearly half of their working hours not working.

The order-batching mathematical model ((10.11)–(10.16)) for the zone-picking strategy when the area size is 2 aisles and 3 aisles is constructed based on the existing data in a similar manner. The optimal solution of each model is obtained. In case (b) where the number of areas is 9; because there is 1 picker in every area, there is a total of 9 pickers and orders numbered 21, 29, 33, 34, and 64 are not assigned to any time windows. Hence, not 14,000 but 13,170 pieces of product can be delivered. The average of capacity utilization rates of the pickers is 99.7 %, and the minimum capacity utilization rate of the pickers is 98.9 %.

In case (c) where the number of areas is 6, because there is 1 picker in every area, there is a total of 6 pickers. Orders numbered 7, 8, 9, 12, 13, 21, 25, 28, 30, 31, 33, 34, 39, 41, 43, 47, 49, 50, 51, 53, 54, 57, 59, 61, 63, and 64 are not assigned to any time windows, and 8400 pieces of product can be delivered. The average of capacity utilization rates of the pickers is 99.7 %, and the minimum capacity utilization rate of the pickers is 99.4 %. The reason why 5600 items cannot be picked, despite the fact that the capacity utilization rate of the pickers is very high, is the insufficient number of pickers and workload imbalance.

10.5.3 Evaluation and Selection

(i) *Calculate the total cost for each alternative*

Five alternative order-picking strategies are listed as follows:

- Alternative 1: pick-and-sort batch-picking strategy with 11_pickers;
- Alternative 2: pick-and-sort batch-picking strategy with 12_pickers;
- Alternative 3: synchronized zone-picking strategy where the area size is 1 aisle (18 zones);
- Alternative 4: synchronized zone-picking strategy where the area size is 2 aisle (9 zones);
- Alternative 5: synchronized zone-picking strategy where the area size is 3 aisle (6 zones);

For each alternative, the annualized cost of a picker, $\boldsymbol{C_p}$, is exactly the same and calculated by multiplying the average monthly labor cost by 12 months. Hence, an employee's annual cost is 3120 TRY * 12 months = 37,440 TRY/year.

For each alternative, the annual depreciation value of a pick cart is 24 TRY and that of each capacity is 250 products. The annual depreciation cost of a handheld barcode scanner terminal is 300 TRY.

Table 10.10 **a** Pickers' workload and capacity utilizations according to zone-picking strategy with 18 areas with $\boldsymbol{M^z} = 770$. **b** Pickers' workload and capacity utilizations according to zone-picking strategy with 9 areas with $\boldsymbol{M^z} = 734$. **c** Pickers' workload and capacity utilizations according to zone-picking strategy with 6 areas with $\boldsymbol{M^z} = 702$

(a)				(b)				(c)			
Picker	Time window	Assigned total product quantity	Capacity utilization (%)	Picker	Time window	Assigned total product quantity	Capacity utilization (%)	Picker	Time window	Assigned total product quantity	Capacity utilization (%)
1	1	595	77.3	1	1	733	99.9	1	1	702	100.0
1	2	183	23.8	1	2	733	99.9	1	2	700	99.7
2	1	595	77.3	2	1	734	100.0	2	1	699	99.6
2	2	183	23.8	2	2	731	99.6	2	2	702	100.0
3	1	595	77.3	3	1	731	99.6	3	1	698	99.4
3	2	182	23.6	3	2	734	100.0	3	2	702	100.0
4	1	596	77.4	4	1	732	99.7	4	1	702	100.0
4	2	182	23.6	4	2	730	99.5	4	2	702	100.0
5	1	596	77.4	5	1	734	100.0	5	1	700	99.7
5	2	181	23.5	5	2	729	99.3	5	2	698	99.4
6	1	597	77.5	6	1	734	100.0	6	1	699	99.6
6	2	181	23.5	6	2	731	99.6	6	2	696	99.1
7	1	595	77.3	7	1	734	100.0				
7	2	183	23.8	7	2	732	99.7				
8	1	595	77.3	8	1	734	100.0				
8	2	183	23.8	8	2	724	98.6				
9	1	597	77.5	9	1	734	100.0				
9	2	181	23.5	9	2	726	98.9				
10	1	596	77.4								

(continued)

Table 10.10 (continued)

(a)				(b)				(c)			
Picker	Time window	Assigned total product quantity	Capacity utilization (%)	Picker	Time window	Assigned total product quantity	Capacity utilization (%)	Picker	Time window	Assigned total product quantity	Capacity utilization (%)
10	2	181	23.5								
11	1	594	77.1								
11	2	183	23.8								
12	1	595	77.3								
12	2	183	23.8								
13	1	593	77.0								
13	2	185	24.0								
14	1	593	77.0								
14	2	185	24.0								
15	1	592	76.9								
15	2	186	24.2								
16	1	592	76.9								
16	2	186	24.2								
17	1	592	76.9								
17	2	186	24.2								
18	1	592	76.9								
18	2	186	24.2								
Average			50.5				99.7				99.7

For each alternative, the overtime labor cost per hour is 19.5 TRY per picker. Based on the performed time study, it was found that the picking operation for 1 item takes 0.01 h. The picking cost of a product with overtime is calculated as 0.01 * 19.5 ≅ 0.20 TRY per item. Because there are 6 working days in a week, the annual overtime cost, C_u, is calculated as 0.20 TRY * 6 days * 52 weeks = 62.4 TRY per item.

Cost Calculation of Alternatives 1 and 2

The cost components of the logical model are presented in (2).

The required number of pickers for Alternative 1, $\boldsymbol{P^{b}_{\mathrm{alt.1}}}$, and that for Alternative 2, $\boldsymbol{P^{b}_{\mathrm{alt.2}}}$, were calculated in step 16 as 11 pickers and 12 pickers, respectively. Hence, $\boldsymbol{P^{b}_{\mathrm{alt.1}} C_{p}} = 11 * 37{,}440$ TRY for Alternative 1 and $\boldsymbol{P^{b}_{\mathrm{alt.2}} C_{p}} = 12 * 37{,}440$ TRY for Alternative 2.

The annual equipment cost, $\boldsymbol{C^{b}_{e}}$, is calculated considering the required number of picking vehicles and handheld terminals for each alternative. In Tables 10.11 and 10.12, the calculation of the required number of picking vehicles is given for the corresponding alternatives. For the calculation of number of pick cart for a picker in

Table 10.11 Number of needed pick cart for Alternative 1

Picker	Time window	Assigned total product quantity	Number of needed pick cart
1	1	570	3
1	2	600	3
2	1	600	3
2	2	520	3
3	1	620	3
3	2	650	3
4	1	520	3
4	2	580	3
5	1	530	3
5	2	630	3
6	1	640	3
6	2	560	3
7	1	650	3
7	2	530	3
8	1	590	3
8	2	600	3
9	1	550	3
9	2	620	3
10	1	640	3
10	2	580	3
11	1	620	3
11	2	450	2
		Total	65

Table 10.12 Number of needed pick cart for Alternative 2

Picker	Time window	Assigned total product quantity	Number of needed pick cart
1	1	630	3
1	2	650	3
2	1	650	3
2	2	650	3
3	1	590	3
3	2	610	3
4	1	610	3
4	2	610	3
5	1	580	3
5	2	620	3
6	1	630	3
6	2	490	2
7	1	640	3
7	2	610	3
8	1	650	3
8	2	620	3
9	1	530	3
9	2	500	2
10	1	640	3
10	2	650	3
11	1	250	1
11	2	380	2
12	1	420	2
12	2	640	3
		Total	66

a time window, the assigned total product quantity is divided by the pick cart capacity (250 products). For example, if the assigned total product quantity is 570, then the required number of pick cart is (570/250) $= 2.28 \cong 3$.

Pick carts used in the time window 1 cannot be reused in time window 2 because pick carts are waiting for the sorting operation until the beginning of the time window 2.

The total number of picking vehicles required for the respective alternatives is 65 and 66. The annual cost of picking vehicles is calculated as 24 (TRY/a picking vehicle) * 65 = 1560 TRY and 24 (TRY/a picking vehicle) * 66 = 1584 TRY for respective cases. When there are 11 pickers and each picker uses 1 handheld terminal, the total cost of handheld terminals is 300 TRY * 11 = 3300 TRY. In this case, the total equipment cost $C^{b}_{ealt.1}$, is calculated as 1560 + 3300 = 4860 TRY. When there are 12 pickers and each picker uses 1 handheld terminal, the total cost of handheld terminals is 300 TRY * 12 = 3600 TRY. In that case, the total equipment cost $C^{b}_{ealt.2}$, is calculated as 1584 + 3600 = 5184 TRY.

The number of items in orders unfulfilled due to workload imbalance in a day in batch picking $\boldsymbol{U^b}$ is calculated based on Eq. (10.17) for Alternatives 1 and 2, respectively. For Alternative 1, the number of products picked is 12,850, and 14,000 pieces of product is required to be fulfilled; therefore, the $U^b_{alt.1}$ value is 1150. For Alternative 2, the number of products picked is 13850, and 14000 pieces of product are required to be fulfilled; therefore, the $U^b_{alt.2}$ value is 150.

The sums of all cost components are calculated for Alternatives 1 and 2 as follows.

$$\begin{aligned}\text{Total cost of alternative 1}\left(C^b_{alt.1}\right) &= P^b_{alt.1}C_p + C^b_{ealt.1} + U^b_{alt.1}C_u \\ &= (11 * 37{,}440) + 4860 + (1150 * 62.4)\text{TRY} \\ &= 488{,}460\ \text{TRY}\end{aligned} \quad (10.19)$$

$$\begin{aligned}\text{Total cost of alternative 2}(C^b_{alt.2}) &= P^b_{alt.2}C_p + C^b_{ealt.2} + U^b_{alt.2}C_u \\ &= (12 * 37{,}440) + 5184 + (150 * 62.4)\text{TRY} \\ &= 463{,}824\ \text{TRY}\end{aligned} \quad (10.20)$$

Cost Calculation of Alternative 3, Alternative 4, and Alternative 5

The cost components of the logical model are presented in (7).

The required picker numbers for alternative 3 $\boldsymbol{P^z_{alt.3}}$, for alternative 4 $\boldsymbol{P^z_{alt.4}}$, and for alternative 5 $\boldsymbol{P^z_{alt.5}}$, were calculated in step 16 as 18 pickers, 9 pickers, and 6 pickers, respectively.

The annual equipment cost, C^z_e, is calculated considering the required number of picking vehicles and handheld terminals. The total numbers of required pick carts are calculated as 72, 54, and 36 for Alternatives 3, 4, and 5, respectively. The calculations are mentioned in Tables 10.13, 10.14, and 10.15. For Alternative 3, the annual cost of the pick cart is 24 * 72 = 1728 TRY. Because there are 18 pickers, and each picker uses 1 handheld terminal, the total cost of the handheld terminal is 300 * 18 = 5400 TRY. The total equipment cost, $C^z_{ealt.3}$, is calculated as 1728 + 5400 = 7128 TRY. For Alternative 4, the annual cost of the pick cart is 24 * 54 = 1296 TRY. As there are 9 pickers and each picker uses 1 handheld terminal, the cost of handheld terminal is 300 * 9 = 2700 TRY. The total equipment cost $C^z_{ealt.4}$, is calculated as 1296 + 2700 = 3996 TRY. For Alternative 5, the annual cost of the pick cart is 24 * 36 = 864 TRY. Because there are 6 pickers, and each picker uses 1 handheld terminal, the total cost of handheld terminals is 300 * 6 = 1800 TRY. The total equipment cost, $C^z_{ealt.5}$, is calculated as 864 + 1800 = 2664 TRY

In Table 10.15, the calculation of the required number of picking vehicles has been given.

The expected number of items in orders unfulfilled due to workload imbalance in a day in zone picking $\boldsymbol{U^z}$ is calculated based on Eq. (10.18) for Alternatives 3, 4, and 5, respectively.

Table 10.13 Number of needed pick cart for Alternative 3

Picker	Time window	Assigned total product quantity	Number of needed pick cart
1	1	595	3
1	2	183	1
2	1	595	3
2	2	183	1
3	1	595	3
3	2	182	1
4	1	596	3
4	2	182	1
5	1	596	3
5	2	181	1
6	1	597	3
6	2	181	1
7	1	595	3
7	2	183	1
8	1	595	3
8	2	183	1
9	1	597	3
9	2	181	1
10	1	596	3
10	2	181	1
11	1	594	3
11	2	183	1
12	1	595	3
12	2	183	1
13	1	593	3
13	2	185	1
14	1	593	3
14	2	185	1
15	1	592	3
15	2	186	1
16	1	592	3
16	2	186	1
17	1	592	3
17	2	186	1
18	1	592	3
18	2	186	1
		Total	72

Table 10.14 Number of needed pick cart for Alternative 4

Picker	Time window	Assigned total product quantity	Number of needed pick cart
1	1	733	3
1	2	733	3
2	1	734	3
2	2	731	3
3	1	731	3
3	2	734	3
4	1	732	3
4	2	730	3
5	1	734	3
5	2	729	3
6	1	734	3
6	2	731	3
7	1	734	3
7	2	732	3
8	1	734	3
8	2	724	3
9	1	734	3
9	2	726	3
		Total	**54**

Table 10.15 Number of needed pick cart for Alternative 5

Picker	Time window	Assigned total product quantity	Number of needed pick cart
1	1	702	3
1	2	700	3
2	1	699	3
2	2	702	3
3	1	698	3
3	2	702	3
4	1	702	3
4	2	702	3
5	1	700	3
5	2	698	3
6	1	699	3
6	2	696	3
		Total	**36**

For Alternative 3, the number of products picked is 14,000 and orders are fulfilled wholly; therefore, the $U^{z}_{\text{alt.3}}$ value is zero. For Alternative 4, the number of products picked is 13,170, and 14,000 pieces of product is required to be fulfilled; therefore, the $U^{z}_{\text{alt.4}}$ value is 830. For Alternative 5, the number of products picked is

Table 10.16 The comparison of alternative order-picking systems based on the total cost

Alternative	Annual cost (TRY)	Number of pickers	Zone size	Average capacity usage rate of pickers (%)
Alternative 1	488,460	11	–	89.4
Alternative 2	463,824	12	–	88.4
Alternative 3	681,048	18	1 Aisle	50.5
Alternative 4	**392,748**	**9**	**2 Aisles**	**99.7**
Alternative 5	576,744	6	3 Aisles	99.7

8400 and 14,000 pieces of product is required to be fulfilled; therefore, the $U^{z}_{\text{alt.5}}$ value is 5600.

The annual overtime cost, $\boldsymbol{C_u}$, is 62.4 TRY per unit product to be picked.

The sums of all cost components calculated for Alternatives 3, 4, and 5 are as follows.

$$\begin{aligned}\text{Total cost of alternative 3 } (C^{z}_{\text{alt.3}}) &= P^{z}_{\text{alt.3}}C_{\text{P}} + C^{z}_{\text{ealt.3}} + U^{z}_{\text{alt.3}}C_{\text{u}} \\ &= (18 * 37{,}440) + 7128 + (0 * 62.4)\text{TRY} \\ &= 681{,}048\,\text{TRY}\end{aligned} \tag{10.21}$$

$$\begin{aligned}\text{Total cost of alternative 4 } (C^{z}_{\text{alt.4}}) &= P^{z}_{\text{alt.4}}C_{\text{P}} + C^{z}_{\text{ealt.4}} + U^{z}_{\text{alt.4}}C_{\text{u}} \\ &= (9 * 37{,}440) + 3996 + (830 * 62.4)\text{TRY} \\ &= 392{,}748\,\text{TRY}\end{aligned} \tag{10.22}$$

$$\begin{aligned}\text{Total cost of alternative 5 } (C^{z}_{\text{alt.5}}) &= P^{z}_{\text{alt.5}}C_{\text{P}} + C^{z}_{\text{ealt.5}} + U^{z}_{\text{alt.5}}C_{\text{u}} \\ &= (6 * 37{,}440) + 2664 + (5660 * 62.4)\text{TRY} \\ &= 576{,}744\,\text{TRY}\end{aligned} \tag{10.23}$$

(ii) *Select the order-picking strategy which has the minimum cost and apply the following orders*

All alternatives with their annual costs are listed in Table 10.16. The zone-picking strategy with two-aisle size has the lowest cost among strategies, and the capacity utilization rates of pickers are also sufficient compared to the other alternatives' utilization rates of pickers. Thereafter, this strategy will be implemented for the next orders.

10.6 Conclusion

In today's competitive environment, warehouse operations play a critical role in maintaining cost effectiveness and service quality in a supply chain. Order picking is the operation with the highest priority due to its high contribution to the total warehouse operating expenses. In this study, an order-picking strategy selection problem in a 3PL firm, which serves a fashion retailing company in Turkey, is considered and solved. The 3PL firm has many customers from different sectors. The customer, whose orders are examined in this study, is primarily engaged in retail sales in the fashion industry.

In this paper, an order-picking system for the customer's fashion products was designed. The proposed flow for the design process consists of strategic decisions, operational implementation, and ES sections. At the operational level of the flow, the pick-and-sort batch-picking strategy with two different levels for the picker number and the synchronized zone-picking strategy with three different zone sizes are utilized as five alternative order-picking strategies based on the considered problem. For each alternative case, the corresponding mathematical models of the order-batching problem are solved to maximize the total number of items fulfilled. The derived optimal solutions of the mathematical models and the rest of the data belonging to the real system are utilized as inputs for the cost centric evaluation process. The best order-picking strategy for the firm is determined as the synchronized zone-picking strategy with a two-aisle zone.

As a remark, the reported literature review revealed that there is no a real-life application study that both proposes a process flow for the design of order-picking systems with detailed implementation steps and utilizes these steps simultaneously. The current study contributes to the literature by incorporating a mathematical modeling based analytical approach into the proposed process flow to determine the optimal order-picking strategy by cost centric evaluation. The process flow can be adapted to any order-picking design and selection problem with slight modifications according to the requirements of the considered system. Furthermore, this study would offer good guidance for warehouse managers to execute the implementation steps of the proposed design flow for real-life applications. Future research can be done by relaxing some model assumptions and trying to apply the analytical models for exploring similar problems in other industrial settings.

References

Accorsi R, Manzini R, Bortolini M (2012) A hierarchical procedure for storage allocation and assignment within an order-picking system. A case study. Int J Logist Res Appl 15(6):351–364

Bozer YA, Sharp GP (1985) An empirical evaluation of a general purpose automated order accumulation and sortation system used in batch picking. Mater Flow 2(2):111–131

Bozer YA, Quiroz MA, Sharp GP (1988) An evaluation of alternative control strategies and design issues for automated order accumulation and sortation systems. Mater Flow 4:265–282

Cagliano CA, DeMarco A, Rafele C, Volpe S (2011) Using system dynamics in warehouse management: a fast-fashion case study. J Manuf Technol Manage 22(2):171–188

Caro F, Gallien J (2015) Inventory management of a fast-fashion retail network. Oper Res 58(2):257–273

Chen CM, Gong Y, DeKoster R, Nunen JAEE (2010) A flexible evaluative framework for order picking systems. Prod Oper Manage 19:70–82

Choi TM (2014) Fashion retail supply chain management—a systems optimization approach. CRC Press, London, UK
Choi TM, Sethi S (2010) Innovative quick response programs: a review. Int J Prod Econ 127(1):1–12
Choy KL, Lam HY, Lin C, Lee CKH (2013) A hybrid decision support system for storage location assignment in the fast-fashion industry. In: Proceedings of technology management in the IT-Driven Services (PICMET), pp 468–473
Christopher M, Peck H (1997) Managing logistics in fashion markets. Int J Logist Manage 8(2):63–74
Dallari F, Marchet G, Melacini M (2009) Design of order picking system. Int J Adv Manuf Technol 42:1–12
DeKoster R, Le-Duc T, Roodbergen KJ (2007) Design and control of warehouse order picking: a literature review. Eur J Oper Res 182:481–501
Ghiani G, Laporte G, Musmanno R (2013) Introduction to logistics systems management, 2nd edn. Wiley, Chichester
Gu J, Goetschalckx M, McGinnis LF (2007) Research on warehouse operation: a comprehensive review. Eur J Oper Res 177:1–21
Hu ZH, Zhao Y, Choi TM (2013) Vehicle routing problem for fashion supply chains with cross-docking. In: Mathematical problems in engineering
Hwang H, Oh YH, Lee YK (2004) An evaluation of routing policies for order-picking operations in low-level picker-to-part system. Int J Prod Res 42:3873–3889
Lin CH, Lu IY (1999) The procedure of determining the order picking strategies in distribution center. Int J Prod Econ 60–61:301–307
Lodree EJ, Geiger CD, Jiang X (2009) Taxonomy for integrating scheduling theory and human factors: review and research opportunuties. Int J Ind Ergon 39:39–51
Mehrjoo M, Pasek ZJ (2014) Impact of product variety on supply chain in fast fashion apparel industry, variety management in manufacturing. In: Proceedings of the 47th CIRP conference on manufacturing systems procedia CIRP 17, pp 296–301
Melacini M, Perotti S, Tumino A (2011) Development of a framework for pick-and-pass order picking system design. Int J Adv Manuf Technol 53:841–854
Meller RD (1997) Optimal order-to-lane assignments in an order accumulation/sortation system. IIE Trans 29:293–301
Meneghetti A (2013) Exploiting fashion features for floor storage systems in the shoe industry. Int J Eng Bus Manage 5:1–12
Parikh PJ, Meller RD (2008) Selecting between batch and zone order picking strategies in a distribution center. Transp Res Part E 44:696–719
Petersen CG (2000) An evaluation of order picking policies for mail order companies. Prod Oper Manage 9(4)
Petersen CG, Aase G (2004) A comparison of picking, storage, and routing policies in manual order picking. Int J Prod Econ 92:11–19
Richards G (2011) Warehouse management—a complete guide to improving efficiency and minimizing costs in the modern warehouse. Kogan Page Limited, UK
Roodbergen KJ (2012) Storage assignment for order picking in multiple-block warehouses, Chap 7. In: Manzini R (ed) Warehousing in the global supply chain. Advanced models, tools and applications for storage systems. Springer, London, pp 373–396
Roodbergen KJ, DeKoster R (2001) Routing methods for warehouses with multiple cross aisles. Int J Prod Res 39(9):1865–1883
Rouwenhorst B, Reuter B, Stockrahm V, Houtum GJ, Mantel RJ, Zijm WHM (2000) Warehouse design and control: framework and literature review. Eur J Oper Res 122:515–533
Shen C, Wu Y, Zhou C (2001) Selecting between sequential zoning and simultaneous zoning for pickertoparts order picking system based on order cluster and genetic algorithm. Chin J Mech Eng 24:1–9
Thompkins JA, White JA, Bozer YA, Frazelle EH, Tanchoco JMA (2003) Facilities planning. Wiley, Hoboken

Yoon CS, Sharp GP (1995) Example application of the cognitive design procedure for an order pick system: case study. Eur J Oper Res 87:223–246

Yoon CS, Sharp GP (1996) A structured procedure for analysis and design of order pick systems. IIE Trans 28:379–389

Yu M, DeKoster R (2009) The impact of order batching and picking area zoning on order picking system performance. Eur J Oper Res 198:480–490

Chapter 11
Fashion Supply Chain Inventory Optimization Models with Service Level and Lead Time Considerations

Guo Li, Yu-chen Kang and Xu Guan

Abstract We study in this paper the optimal inventory models in fashion supply chains with a controllable lead time and a service-level constraint. First, we review the analytical inventory model which assumes that the lead time can be divided into n parts, and the cost of compressing lead time has a linear relation with the lead time. Then, we examine the optimal quantity decisions in the fashion supply chain. Furthermore, extensive numerical analyses are conducted to generate important insights into the problem. Future research directions are discussed.

Keywords Service level · Controllable lead time · Fashion supply chains

11.1 Introduction

Fashion products have some specific features such as short product life cycle, volatile, volatile demand, and great product variety (with many SKUs). Therefore, an efficient supply chain is critically important to ensure the proper management of

G. Li (✉) · Y. Kang
School of Management and Economics, Beijing Institute of Technology,
No. 5 Zhongguancun South Street, Haidian District, Beijing 100081, China
e-mail: liguo@bit.edu.cn

Y. Kang
e-mail: 471570626@qq.com

G. Li · Y. Kang
Center for Energy and Environmental Policy Research,
Beijing Institute of Technology, Beijing 100081, China

X. Guan
School of Economic and Management, Wuhan University,
No. 299 Bayi Road, Hongshan District, Wuhan, Hubei 430072, China
e-mail: gavinguan@whu.edu.cn

T.-M. Choi (ed.), *Analytical Modeling Research in Fashion Business*,
Springer Series in Fashion Business, DOI 10.1007/978-981-10-1014-9_11

them (Şen 2008). In the fashion industry, some successful fashion retailers called the fast fashion retailers are famous for their speed in offering a large variety of products to the market. In particular, a leading brand Zara can efficiently complete the whole design–production–distribution cycle in only about 15 days. All of these rely on an efficient supply chain and inventory management system. In the past, much effort was made in examining the optimal fashion supply chain decisions with the considerations of demand variations, pricing, and other operational issues (Wang et al. 2012; Li et al. 2014a; Peng and Zhou 2010; Chiang and Benton 1994; Treleven and Schweikhart 1988). However, lead time management and inventory service level are definitely two crucial and fundamental aspects for proper inventory management in fashion supply chain systems.

In inventory management, it is well known that customer service can be reflected by the inventory service level. A high inventory service level implies a high level of customer service which yields high customer satisfaction (Innis and La Londe 1994). Service level, also regarded as customer satisfaction, refers to the quality of service offered by companies. Service level plays an important role in companies because of its tremendous effect on the development of a firm. For fashion supply chains, service level is also important. This paper mainly focuses on an optimal inventory model which subjects to shorter lead time with service-level constraint. An increasing number of managers realize the benefit of reducing lead time, which has attracted a number of scholars to research on this topic. Therefore, the growing effort of optimizing fashion supply chains is added by analyzing order and production decisions under service-level constraint and controllable lead time.

Yet the optimal decision of fashion supply chains under service-level constraint and controllable lead time has never been adequately investigated. The optimal back-order decision and lost sales problem have a close bearing with service-level constraint and controllable lead time (Chu et al. 2005). Speed is becoming a strategic competitive weapon, and purchasing optimal lead time has gradually attracted the attention of enterprises and researchers. Thus, studying the order and production decisions of the fashion supply chain with service-level constraint and controllable lead time is of high significance. This paper first reviews the analytical models of a related paper (Li et al. 2014b) and examines the optimal ordering, production quantity in the fashion supply chain with service-level constraint, and controllable lead time. With extensive numerical analyses, this paper generates important insights regarding the inventory management problems for fashion companies. Future research directions are also discussed.

This paper is organized as follows. Section 11.2 reviews the relevant literature. In Sect. 11.3, we review the related model. In Sect. 11.4, numerical analysis is conducted to present the effects of parameters on the optimal solution. Section 11.5 concludes the paper with a discussion on future research.

11.2 Literature Review

Due to global competition, fashion companies are aiming to achieve a more effective supply chain in order to provide quality products with lower expenses, at the right time and in the right quantity to the market. This paper is built upon three streams of research in the literature: (a) inventory management with service-level constraint, (b) lead time management, and (c) fashion inventory management. We review them as follows.

Instead of using a shortage cost term in the objective function, Jha and Shanker (2009) use a service-level constraint to formulate a optimal model. In today's competitive environment, service level and lead time need to be constantly improved to achieve optimal performance of a supply chain. Tersine (1982) indicates that lead time usually includes the following components: order preparation, order transit, supplier lead time, delivery time, and setup time (i.e., preparation time for availability). In practice, lead time can be reduced by an added crashing cost, which is controllable in other words. A short lead time can improve customer service levels, reduce stock costs, and ensure safety stock, among others. Thus, lead time has attracted the attention of scholars, and numerous inventory models have considered lead time and service level (Ben-Daya and Raouf 1994; Ouyang and Wu 1997; Hariga and Ben-Daya 1999; Ben-Daya and Hariga 2003; Chu et al. 2005; Li et al. 2013, 2014c). In this section, we review related literature on two aspects: lead time and service level.

Various models and mechanisms have been developed to provide a more effective and scientific way of compressing controllable lead time. Liao and Shyu (1991) are among the first ones to assume that lead time is negotiable and can be decomposed into several components, with each component having a different piecewise linear crash cost function for lead time reduction. Subsequently, an increasing number of inventory models for the lead time reduction issue have been developed. For example, Ben and Raouf (1994) extend the model of Liao and Shyu by viewing both lead time and order quantity as decision variables and then proposed a management and control strategy for lead time on account of these variables. Based on this strategy, Ouyang et al. (1996) further consider the circumstances of back orders and stock out. Moon and Choi (1998) and Lan et al. (1999) find individual optimal order quantity and optimal lead time for a mixed inventory model, and they develop a simplified solution under a continuous inventory system. Ouyang et al. (2002) also attempt to derive a continuous review inventory model with defective items, where the order quantity, reorder point, and lead time are regarded as decision variables. They initially assume that the lead time demand follows a normal distribution and then that the mean and variance of lead time demand are known. Ouyang and Wu (1998), for the first time, integrate the distribution-free condition into the inventory model. They loosen the assumption about the cumulative lead time distribution demand while developing the minimized distribution-free procedure to determine the optimal lead time and optimal order quantity. Ouyang et al. (2006) then investigate the effect of investing in quality improvement and lead time reduction on the integrated vendor–buyer inventory model with partial back orders and apply the minimized distribution-free procedure to solve the problem. Jha and Kripa (2013) extend the integrated vendor–buyer inventory model to single-vendor, multi-buyer-integrated production inventory model. In 2013, a breakthrough on the cost function of compressing lead time is found. Li et al. (2013) propose that the unit cost of compressing lead time in most periods includes the fixed and the variable parts and that these variable parts are sensitive to order quantity. More reasonable lead time unit compressing cost should be segmented and be made sensitive to order quantity and productivity (Ouyang et al. 2004; Lee et al. 2007; Rho and Yu 1998).

In most papers, a discrete distribution for the lead time is developed from the historical data. Almost all inventory models in the literature assume that lead time is prescribed and not subject to control (Eppen and Martin 1988; Liao and Shyu 1991). Exponential distribution is widely used to model many time-oriented variables, such as waiting time or lead time between occurrences of events and during the lifetime of electrical, mechanical devices, or the fashion industry (Sajadieh and Akbari Jokar 2009). However, this paper assumes that the lead time has n mutually independent components, especially with the particularity of fashion supply chains, because a longer lead time corresponds to a smaller unit cost of compressing lead time in clothing productive processing. Hence, the unit lead time of compressing cost should be shorter than regular lead time.

Most of the previously mentioned literature introduces the theory of controllable lead time into the inventory model, but a few of them simultaneously consider service-level constraint. Larson (1998), who first explains the concept of the service-level agreement in IT service provision, introduces the state of outsourced service provision. In his paper, he provides the structure of good service-level agreement. Jain et al. (2002) further mention that the Internet service has an infrastructure that involves cooperation between the organization and the system. The service level in IP network agreements quantifies the level in nature, such that the metrics used to define the quality of service are included. IP networks are also shown to necessarily improve its service level. For meeting the objective of the service level, Chen et al. (2003) establish a method of decreasing the total cost of employee operation hours in medical care for employees to meet the service-level objective. This method is a fair evaluation of the service level. Relative service efficiency is also emphasized as an indicator to measure the service level in this paper. The indicator relates to the relative service efficiency of a general hospital, but not to the categorical performance result. Thus, a fairer evaluation of the service level is established. Liu and Ma (2005) study the effect of service level on the acceptance of application service-oriented medical records and demonstrate service level as "the most important criterion in evaluating application services." Most importantly, the study by Liu et al. investigates the effect of service-level perception on the willingness of healthcare workers to use application service-oriented medical records. For the service level in the supply chain, service-level perception is often introduced into the inventory model because decision makers realize that stock out and customer loss, among others, may result in without this factor, which is especially significant for fashion supply chains. Ouyang and Wu (1997) suggest that a service-level constraint, which is a bounded stock-out level per cycle, be added to the model instead of having a stock-out term in the objective function. Chiu et al. (2014) consider a market based on service-level decision on a particular common service product, consequently finding one service provider choosing to deviate from the equilibrium service level with the goal of improving market share. Hence, service-level constraint is necessary in considering the stock-out problem with controllable lead time, but only a few studies have included this factor.

Lastly, Choi et al. (2013) indicated that service operations are critical in modern businesses. Therefore, in analyzing the supply chains of the fashion industry, this paper considers service level and the optimal cost of having a controllable lead time. Furthermore, with the fast development of fashion industry economy, the efficiency of fashion supply chain has become the focus and to study fashion supply chain management become a top priority (Hinkka et al. 2015). Brun and Castelli (2008) develop a portfolio model to study the fashion supply chain strategy. To explore how high-performing companies in the fashion industry align their supply chain strategy with their competitive priorities, Kim (2013) adopts in-depth interview with top managers in the companies and analyses the competitive priorities, so as to study the supply chain strategy. Our paper also analyses the fashion inventory management and is therefore related to the above studies. However, we study a different supply chain with service-level constraint and controlled lead time. We find that the supply chain will be improved in some extent with service-level constraint and controlled lead time.

11.3 The Analytical Model

11.3.1 Notations

The following notation is employed in the model development and analysis, as shown in Table 11.1.

In the inventory model, we consider a fashion supply chain consisting of a single supplier and a single buyer. The buyer employs a continuous review inventory policy. In other words, whenever the inventory level falls below the reorder point r, a reorder is placed automatically via a computerized system. For the reorder point, it follows the standard formula in the literature: $r = \mu L + k\sigma\sqrt{L}$, where k is the safety factor (Tersine 1982). By definition, the service-level constraint is $\frac{E\left([X-r]^+\right)}{q} \leq \theta$. The lead time L consists of n independent components, among which the shortest duration is t_j and the normal duration is T_j. Assume that the unit compressing costs of lead time is a_i. For a given $L \in \left[L_{i,}L_{i-1}\right]$, the lead time compressing cost per cycle is that $C(L) = a_i(L_{i-1} - L) + \sum_{j=1}^{i-1} a_j\left(T_j - t_j\right)$ where $j = 1, 2, \ldots, i$.

Table 11.1 Instruction for variables

Variable	Definition
μ	The average weekly demand
σ^2	The variance of weekly demand
q	Order quantity of the buyer
L	The length of lead time for the buyer (in weeks)
m	The number of lots in which the product is delivered from the vendor to the buyer in one production cycle, a positive integer
α	The ratio of price discount offered by the vendor, $\alpha \in [0, 1]$
α'	The price discount of unit reorder product
k	Safety coefficient, $k \geq 0$
r	Reorder point of the buyer
θ	Service-level coefficient
$B(r)$	Expected demand shortage at the end of cycle
X	The lead time demand that follows a normal distribution with finite mean μL and a standard deviation $\sigma\sqrt{L}$
P_v	The vendor's unit wholesale price
P	The buyer's unit retail price
P_c	The vendor's production capability per week ($P_c > \mu$)
β	Proportion of customers who are willing to wait for next replenishment $\beta \in [0, 1]$
S	The vendor's setup cost per setup
r_b	Buyer's holding cost rate (per monetary unit invested in inventory) per week
r_v	Vendor's holding cost rate (per monetary unit invested in inventory) per week

11.3.2 Model Formulation

1. **The buyer's inventory model**
   We consider the buyer's expected total cost per week as the sum of purchasing cost, holding cost, shortage cost, and the lead time crash cost: $C_{\mathrm{BT}} = C_{\mathrm{BP}} + C_{\mathrm{BH}} + C_{\mathrm{BS}} + C_{\mathrm{BL}}$. The buyer's expected purchasing cost per week is as follows: $C_{\mathrm{BP}} = P_v\mu$. Following the approach in (Li et al. 2014b), the buyer's expected total cost per week is derived to be the following:

$$C_{\mathrm{BT}} = P_v\mu + r_bP_v\left\{\frac{q}{2} + k\sigma\sqrt{L} + (1-\beta)\sigma\sqrt{L}\psi\right\} + \frac{\mu}{q}P\sigma\sqrt{L}\psi(k)(1-\alpha'\beta) + \frac{\mu}{q}C_L$$
$$= P_v\mu + r_bP_v(\frac{q}{2} + k\sigma\sqrt{L}) + \sigma\sqrt{L}\psi(k)\left[r_bP_v(1-\beta) + \frac{P\mu}{q}(1-\alpha'\beta)\right] + \frac{\mu}{q}C_L$$

2. **The vendor's inventory model**
   Similar to most research dealing with controllable lead time with the same setting for the vendor (e.g., Ouyang et al. 2004; Li et al. 2013), the vendor's expected cost per week is represented as:

$$
\begin{aligned}
C_{\mathrm{VT}} &= \frac{S\mu}{mq} + \frac{r_v P_v q}{2}\left[m\left(1-\frac{\mu}{P_c}\right) - 1 + \frac{2\mu}{P_c}\right] \\
&= \frac{S\mu}{mq} + \frac{r_v P_v q(m-1)}{2} + \frac{(2-m)\mu r_v P_v q}{2P_c}
\end{aligned}
\quad (11.1)
$$

11.3.3 Model Analysis

To obtain the optimal order quantity, optimal production quantity, and optimal lead time with service-level constraint, we can adopt the first-order conditions. For the buyer, it is straightforward to derive that the optimal order quantity is given as follows:

$$
q_{\mathrm{BT}} = \max\left\{\left[\frac{2\mu[\sigma\sqrt{L}\psi(k)p(1-\alpha'\beta)+C_L]}{r_b p_v}\right]^{\frac{1}{2}}, \frac{\sigma}{\theta}\sqrt{L}\psi(k)\right\}, L \in [L_i, L_{i-1}]. \quad (11.2)
$$

For the vendor, its total expected cost is given by Eq. (11.1), and it can be minimized by taking the optimal order quantity and lead time already determined for the buyer and selecting suitable integer value of m. To find the optimal m, we employ the algorithm as shown in Li et al. (2014b).

11.4 Numerical Analysis

To test the feasibility of the model, we consider a fashion company with the following parameters:$\theta = 5\,\%$, $P_v = 60$ USD/unit, $P = 125$ USD/unit, $r_b = 0.25$, $r_v = 0.25$, $P_c = 3200$ units/a, $\beta = 0.7$, $\alpha' = 0.75$, $S = 1340$ USD/time, $\sigma = 20$ units/week, and $\mu = 2400$ units/a. The lead time composition is shown in Table 11.2. Notice that these numbers are selected in a way in which they satisfy the model assumptions and are rather representative of the real-world scenario.

First, we can use the algorithm to get different optimal C^*_{BT} with the different safety factors k as given in Table 11.4. From Table 11.3, we can see that if the safety factor k increases, the optimal buyer's expected total cost per week C^*_{BT} will decrease and increase contrarily.

Figures 11.1 and 11.2 give the description of the effect of q on the joint cost and the buyer's cost. With different service levels, we obtain the variance of the optimal buyer's cost under different values of k from Fig. 11.3. And there is a U-shaped relationship between the buyer's expected total cost per week C_{BT} and the safety coefficient k. The buyer's inventory cost in Fig. 11.3 decreases until k reaches a certain value, and then, the buyer's inventory cost begins to increase, which means

Table 11.2 Lead time constituent data

Lead time composition i	Normal lead time T_i (days)	Shortest lead time t_i (days)	Unit compressing cost a_i (USD/day)
1	20	6	0.1
2	20	6	1.0
3	16	9	5.0

Table 11.3 Fashion inventory optimization with different safety factor k

k	m	$q_{\min}$	$q_{\max}$	q^*	L_i^*	C_{BT}^*
0.00	1	1192	1199	1195	2	154723*
	2	1192	1199	1193	4	155436
0.25	1	1054	1061	1058	3	152149*
	2	1054	1061	1055	4	152158
0.50	2	937	946	943	4	152147
	3	937	946	941	4	151738*
0.75	2	843	853	852	4	151713*
	3	843	853	849	6	151988
1.00	2	767	778	776	6	152478*
1.25	2	706	718	710	6	153276
1.50	1	656	668	665	6	154411*
	2	656	668	657	8	154417

The lead time L_i^* unit is week
$*$ is the optimal solution with the corresponding safety factor

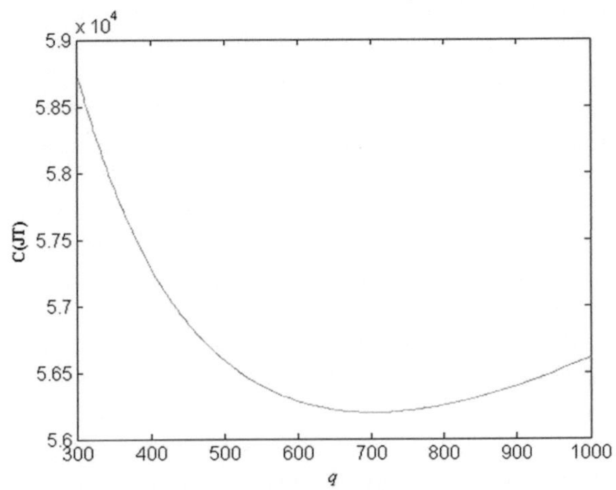

Fig. 11.1 The effect of q on the optimal joint total cost

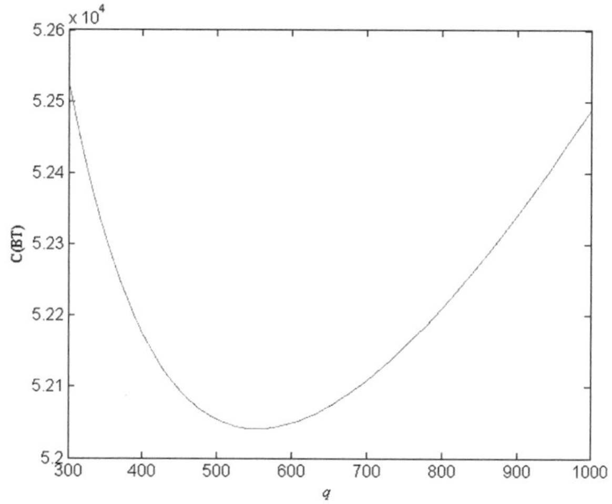

Fig. 11.2 The effect of q on the optimal buyer's cost

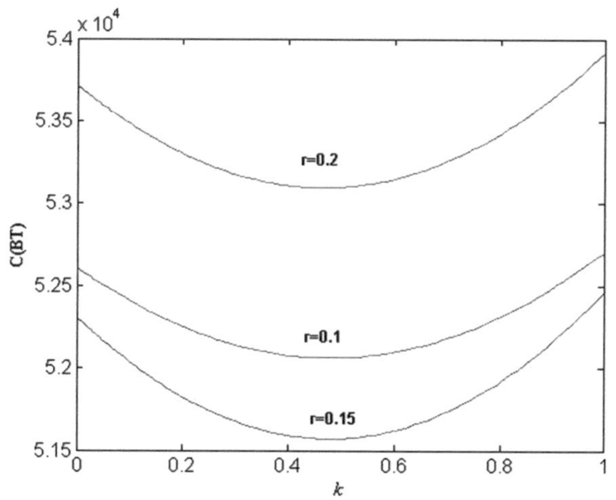

Fig. 11.3 Variance of the optimal buyer's cost under different values of k with different service level

that the buyer is up to the scale and an optimal k making the buyer's inventory cost minimum is found. Figure 11.4 describes the effect of L on q with different m, which implies that the value of q increases when the value of L increases at a constant value of m, which means the improved value of L cannot always increase the order quantity under the constraint of m. The vendor and buyer should consider everything to decide the value of q. From Fig. 11.4, we know it is obvious that the lead time has an optimal value, and the order quantity will reach the maximum.

Table 11.4 shows the effect of holding costs of both the vendor and the buyer on the joint expected total cost per week C_{JT}. Given that the buyer's holding cost rate (per monetary unit invested in inventory) per week r_b and the vendor's holding cost rate (per monetary unit invested in inventory) per week r_v may not be the same in

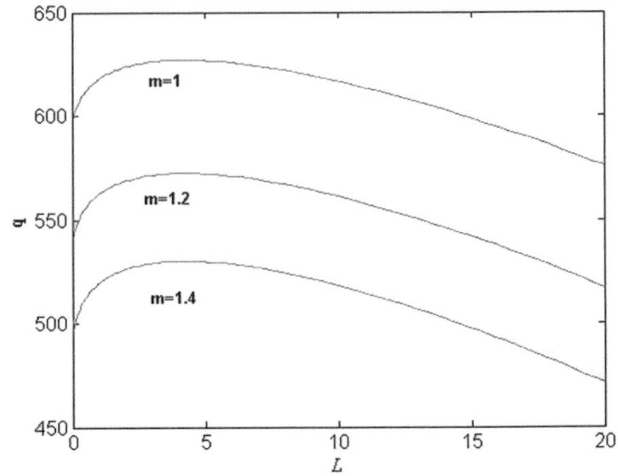

Fig. 11.4 The effect of L on q at different values of m

practice, we analyze the effect of holding cost rate on the joint total expected cost, as shown in Table 11.4. When $r_b > r_v$, the buyer tends to order a small quantity. However, this tendency has a small effect on the vendor. When the buyer has a high holding cost rate, he/she is more likely to order a smaller quantity each time and to lower his/her average inventory level as possible by maintaining higher safety stock. When r_b reaches a certain value, it will have an increasingly less effect on order quantity. However, when $r_b < r_v$, the buyer will order a larger quantity and the vendor slightly changes. Therefore, if the buyer has a smaller holding cost rate, he/she is more likely to order a larger quantity each time to maintain his/her average inventory level. However, small changes in r_v have little effect on product quantity for the vendor. The above discussion explicitly indicates that the vendor prefers to produce in smaller lot size in each production setup so that the vendor could be able to keep his average inventory level as low as possible. Therefore, one may infer that in the practice, the vendor would prefer a low lot size for a low inventory. This argument is also consistent with our observations in the fast fashion industry.

Table 11.4 Optimal results for different values of r_b and r_v (lead time in weeks)

r_b	r_v	m^*	L^*	q^*	C^*_{BT}
0.10	0.10	4	8	653	121140
	0.15	4	8	629	123679
	0.20	3	7	599	134158
0.15	0.10	5	7	570	134233
	0.15	4	7	565	136775
	0.20	4	7	541	136981
0.20	0.10	6	7	540	136981
	0.15	5	6	533	137940
	0.20	4	6	512	141452

Contrary to the single style and big lot size with the traditional fashion, fast fashion companies tend to choose to produce multiple styles and with a low lot size. It is the multiple styles and low lot size that contribute to the low inventory level of fast fashion companies. For the fashion industry, inventory means the losses, because fashion products gain nothing once be stranded in the warehouse.

11.5 Conclusion and Future Research

In this paper, we have reviewed the problem of a two-echelon fashion supply chain consisting of a single vendor and a single buyer with special focal points on the lead time decision under the service-level constraint. Observe that lead time has become increasingly important for fashion supply chains. This is partially driven by the fast fashion business trend. In fact, the length of lead time represents speed and service level. In this study, lead time is controllable, and the lead time has been divided into n parts for the analysis purpose. We have conducted extensively numerical analyses to generate important insights. Notice that the optimal order and production quantities have their boundaries because of the service-level constraint.

In the future, more efforts can be put on generalizing the model to the case with more real-world parameters and decisions such as product pricing. It is also important to consider the issues of supply chain coordination. This paper only examines a single-vendor single-buyer two-echelon fashion supply chain. Extensions can hence be done to explore the more complex supply network with more than two echelons and non-serial structure. Last but not least, this paper assumes that lead time is controllable with certainty. In future research, we can relax this assumption and incorporate the stochastic component into the lead time formulation.

Acknowledgments This work was supported by the National Natural Science Foundation of China (no. 71102174 and 71372019), Beijing Higher Education Young Elite Teacher Project (no. YETP1173), Specialized Research Fund for Doctoral Program of Higher Education of China (no. 20111101120019), and Beijing Philosophy and Social Science Foundation of China (no. 11JGC106).

References

Ben-Daya M, Raouf A (1994) Inventory models involving lead time as decision variable. J Oper Res Soc 45(5):579–582

Ben-Daya M, Hariga M (2003) Lead time reduction in a stochastic inventory system with learning consideration. Int J Prod Res 41(3):571–579

Brun A, Castelli C (2008) Supply chain strategy in the fashion industry: developing a portfolio model depending on product, retail channel and brand. Int J Prod Econ 116(2):169–181

Chen MS, Lin C, Lee CC (2003) Determining employee operation hours in general hospital with expected service level. J Statistics Manage Syst 6(2):241–252

Chiang C, Benton WC (1994) Sole sourcing versus dual sourcing under stochastic demands and lead times. Naval Res Logistics 41(5):609–624

Chiu CH, Choi TM, Li YJ, Xu L (2014) Service competition and service war: a game theoretic analysis. Serv Sci (INFORMS J) 6:63–76

Choi TM, Chow PS, Govindan K (2013) Service optimization and control. Math Prob Eng. doi:10.1155/2013/951376. Article ID 951376

Chu P, Yang KL, Chen PS (2005a) Improved inventory models with service level and lead time. Comput OR 32(2):285–296

Chu P, Yang K, Chen P (2005b) Improved inventory models with service and lead time. Comput Oper Res 32(2):285–296

Eppen GD, Martin RK (1988) Determining safety stock in the presence of stochastic lead time and demand. Manage Sci 34(11):1380–1390

Hariga M, Ben-Daya M (1999) Some stochastic inventory models with deterministic variable lead time. Eur J Oper Res 113(1):42–51

Hinkka V, Häkkinen M, Holmström J, Främling K (2015) Supply chain typology for configuring cost-efficient tracking in fashion logistics. Int J Logistics Manage 26(1):42–60

Innis DE, La Londe BJ (1994) Customer service: the key to customer satisfaction, customer loyalty, and market share. J Bus Logistics 15(1):1–27

Jain G, Singh D, Verma S (2002) Service level agreements in IP networks. Inf Manage Comput Secur 10(4)

Jha JK, Shanker K (2009) Two-echelon supply chain inventory model with controllable lead time and service level constraint. Comput Ind Eng 57(3):1096–1104

Jha JK, Kripa S (2013) Single-vendor multi-buyer integrated production-inventory model with controllable lead time and service level constraints. Appl Math Model 37(4):1753–1767

Kim B (2013) Competitive priorities and supply chain strategy in the fashion industry. Qual Mark Res Int J 16(2):214–242

Lan SP, Chu P, Chung KJ, Wan WJ (1999) A simple method to locate the optimal solution of inventory model with variable lead time. Comput Oper Res 26:599–605

Larson KD (1998) The role of service level agreements in IT service delivery. Inf Manage Comput Secur 6(3)

Lee WC, Wu JW, Lei CL (2007) Computational algorithmic procedure for optimal inventory policy involving ordering cost reduction and backorder discounts when lead time demand is controllable. Appl Math Comput 189(1):186–200

Li G, Liu MQ, Wang ZH, Peng BZ (2013a) Supply Coordination Based on Bonus Policy in Assembly under Uncertain Delivery Time. Chinese J Mech Eng 26(2):293–303

Li G, Guan X, Liu MQ, Shan MY (2013b) Two-echelon inventory model with service level constraint and controllable lead time sensitive to order quantity. J Central S Univ 20(11):3324–3333

Li G, Kang Y, Liu M, Wang Z (2014a) Optimal inventory policies under permissible payment delay in fashion supply chains. Math Probl Eng. Article ID 327131

Li G, Kang YC, Guan X (2014b) Optimal decision for fashion supply chains with service level constraint and controllable lead time. Math Probl Eng. Article ID 758494

Li G, Liu MQ, Guan X, Huang Z (2014c) Collaborative policy of the supply-hub for assemble-to-order systems with delivery uncertainty. Discrete Dyn Nat Soc. doi:10.1155/2014/625812

Liao CJ, Shyu CH (1991a) An analytical determination of lead time with normal demand. Int J Oper Prod Manage 11:72–78

Liao CJ, Shyu CH (1991b) An analytical determination of lead time with normal demand. Int J Oper Prod Manage 11(9):72–78

Liu LP, Ma QX (2005) The impact of service level on the acceptance of application service oriented medical records. Inf Manag 42(8):1121–1135

Moon I, Choi S (1998) A note on lead time and distribution assumptions in continuous reviews inventory models. Comput Oper Res 25:1007–1012

Ouyang LY, Wu KS (1997) Mixture inventory model involving variable lead time with a service level constraint. Comput Oper Res 24(9):875–882
Ouyang LY, Wu KS (1998) A mixture distribution free procedure for mixed inventory model with variable lead time. Int J Prod Econ 56:511–516
Ouyang LY, Yeh NC, Wu KS (1996) Mixture inventory model with backorders and lost sales for variable lead time. J Oper Res Soc 47:829–832
Ouyang LY, Chuang BR, Chang PH (2002) Lead time reduction models with defective items subject to a service level constraint. J Inf Optimization Sci 23(1):133–146
Ouyang LY, Wu KS, Ho CH (2004) Integrated vendor buyer cooperative models with stochastic demand in controllable lead time. Int J Prod Econ 92(3):255–266
Ouyang LY, Wu KS, Ho CH (2006) The single-vendor single-buyer integrated inventory problem with quality improvement and lead time reduction-minimax distribution-free approach. Asia-Pac J Oper Res 23(3):407–424
Peng HJ, Zhou MH (2010) Production and ordering decision in supply chain with uncertainty in two-echelon yields and demand. J Syst Eng 25(5):622–628 (In Chinese)
Rho BH, Yu YM (1998) A comparative study on the structural relationships of manufacturing practices, lead time and productivity in Japan and Korea. J Oper Manage 16(2):257–270
Sajadieh MS, Jokar MRA (2009) An integrated vendor-buyer cooperative model under stochastic supply lead-time. Int J Adv Manufact 41:1043–1050
Şen A (2008) The US fashion industry: a supply chain review. Int J Prod Econ 114(2):571–593
Tersine RJ (1982) Principles of inventory and materials management. North-Holland, Amsterdam
Treleven M, Schweikhart SB (1988) A risk/benefit analysis of sourcing strategies: single vs. multiple sourcing. J Oper Manage 7(3–4):93–114
Wang K, Gou QL, Sun JW, Yue XH (2012) Coordination of a fashion and textile supply chain with demand variations. J Syst Sci Syst Eng 21(4):461–479

Chapter 12
An AHP-Based Scheme for Sales Forecasting in the Fashion Industry

Ying Zhang, Chunnan Zhang and Yu Liu

Abstract In this paper, we propose a novel aggregation–disaggregation scheme based on the analytic hierarchy process (AHP) to conduct sales forecasting in the fashion industry, where a stock keeping unit (SKU) is usually characterized by a short product life cycle and low sales volume because of various reasons such as the wide variety of products and the constant changes in fashion trends. To improve the accuracy of sales forecast, an AHP-based scheme is proposed to aggregate historical sales data so that future sales can be predicted with higher certainty, and then disaggregate forecast quantities over SKUs, products, or other entities of interest. The proposed scheme is evaluated by using real data collected from the fashion industry, and the experimental results show that under this scheme, the performance of two popular and well-established time series forecasting techniques, namely "moving average" and "exponential smoothing," can be improved to predict the sales of the products with short history and low sales volumes. To continue this work, future research directions are also discussed.

Keywords Analytic hierarchy process · Sales forecasting · Fashion industry · Pairwise comparison · Supply chain

---

Y. Zhang (✉)
Department of Industrial and Systems Engineering,
The State University of New York at Buffalo, 342 Bell Hall,
Buffalo, NY 14260, USA
e-mail: yz29@buffalo.edu

C. Zhang
Department of Accounting, Temple University, 1801 Liacouras Walk,
Philadelphia, PA 19122, USA
e-mail: chunnan.zhang@temple.edu

Y. Liu
College of Computer Science and Technology,
Wuhan University of Science and Technology, 947 Heping Avenue,
Wuhan, Hubei 430081, China
e-mail: liuyu@wust.edu.cn

T.-M. Choi (ed.), *Analytical Modeling Research in Fashion Business*,
Springer Series in Fashion Business, DOI 10.1007/978-981-10-1014-9_12

12.1 Introduction

Forecasting is a very important topic in academic research and has been widely applied in many fields such as supply chain management (Cui et al. 2015; Schwartz et al. 2009; Jin et al. 2015), reliability engineering (Hu et al. 2011), and semiconductor manufacturing (Chen and Wang 2014; Luo et al. 2015). In reality, many companies with global supply networks suffer from market volatility and supply disruptions, which adversely affect both their short- and long-term profits (Asian and Nie 2014). Thus, forecasting is the preliminary step for many important business decisions such as production planning (Albey et al. 2015), spare parts management (Heinecke et al. 2013), and new product launch strategy (Cui et al. 2011) and hence a key driver to improve supply chain performance (Thomassey 2010).

For a fashion business, advanced statistical approaches and artificial intelligence techniques have been widely used to predict both future sales (Liu et al. 2013; Nenni et al. 2013) and fashion trend (Choi et al. 2012; Yu et al. 2012). Both of these problems are hard to solve because the fashion industry is characterized by short product life cycles, volatile customer demands, tremendous product varieties, and complex supply chains (Sen 2008). Also, the complexity of sales forecasting in the fashion industry is increased by the strong seasonality and the frequently changing market environment. Because of the great impact of forecasting on many aspects of the business such as operational performance (Danese and Kalchschmidt 2011), forecasting problems in fashion business have been studied extensively from many different perspectives in the literature. For example, by introducing advanced artificial intelligence techniques such as neural networks and fuzzy logic, Au et al. (2008) proposed an evolutionary neural networks approach in searching for the ideal network structure for a forecasting system and developed the optimized neural networks structure for the forecasting of apparel sales; Sun et al. (2008) applied a novel neural network technique called extreme learning machine (ELM) to investigate the relationship between sales volumes and some significant factors which affect demand, and Kaya et al. (2014) developed a fuzzy forecast combiner which calculates the final forecast using a weighted average of forecasts generated by independent methods. For different time spans, Du et al. (2015) proposed a multiobjective optimization-based neural network model to tackle the short-term replenishment forecasting problem in fashion industry, and Wong and Guo (2010) developed a hybrid intelligent (HI) model which comprises a data preprocessing component and an HI forecaster to tackle the medium-term fashion sales forecasting problem. Since time efficiency is also very important for the industry practices where data volume is very high, research efforts are also conducted to develop fast forecasting tools. For example, Yu et al. (2011) proposed a new fast forecasting model which employs both the extreme learning machine (ELM) and the traditional statistical methods as a quick and effective tool, and Choi et al. (2014) developed an intelligent forecasting algorithm which combines tools such as the extreme learning machine and the gray model to support operational decisions in fast fashion business.

It is observed that the sales of fashionable products follow unpredictable fashion trends associated with high volatility and strongly depend on many factors such as prices and economic conditions (Ren et al. 2015). Also, different from other traditional products, a product line always comprises a lot of stock keeping units (SKUs), and the sales of those SKUs are usually correlated. Thus, fashion sales are not only influenced by some important factors such as color or price, but also by the sales of correlated items. To study a multidimensional relationship between sales volumes and other influence factors in fashion sales forecasting, Ren et al. (2015) developed a panel data method supported by particle filter and conducted a numerical experiment in terms of item and color categories. Although the method outperforms some other popular statistical and intelligent approaches in the literature, the correlation between items is not closely examined. Thus, it will be of great interest to explore new methods which can shed light on the quantitative relationship between the products under study as well as improve forecast accuracy.

In this paper, we propose a novel method based on the analytic hierarchy process (AHP) to study the quantitative relationship between fashion products, and then use it to improve forecast accuracy by taking advantage of data aggregation. In the fashion industry, even if the aggregate demand can be predicted with some certainty, it is still very difficult to predict how it will be distributed over the many products that are offered because of the low sales volumes of individual SKUs and the significant variation of the demand of the SKUs within a same product line (Mostard et al. 2011). As an effort to use aggregation–disaggregation process, Bruzzone et al. (2013) proposed a forecasting model based on multiple autoregressive algorithms and disaggregation policies. In our scheme, since products can be weighted by their percentages in the aggregate sales volume, it will be easy and straightforward to make a division over products that are offered for sales. Moreover, since the AHP-based scheme is a framework which is designed to aggregate historical observations and disaggregate forecast volumes properly, it can be applied to most, if not all, existing approaches whose effectiveness can be improved if there are more stable data for analysis.

The rest of this paper is organized as follows. First, we provide a brief introduction to the AHP and propose the forecasting aggregation and disaggregation scheme with all necessary details in Sect. 12.2. Then, we present numerical experiments and discuss findings in Sect. 12.3. Last, we conclude this paper in Sect. 12.4.

12.2 AHP-Based Scheme for Data Aggregation and Disaggregation in Fashion Sales Forecasting

12.2.1 Introduction to the AHP

The AHP was first introduced by Saaty (1980) as a structured technique for complex decisions. Since then, it has been a popular approach among various multiple-criteria decision-making (MCDM) techniques proposed in the literature and applied in a wide variety of problems which comprise planning, selecting alternatives, allocating resources, and resolving conflicts (Subramanian and Ramanathan 2012). The AHP presents an effective way to combine subjective human knowledge with objective analysis and provides a solid framework to structure a problem and evaluate alternative solutions. In an AHP application, a decision problem will be decomposed into a hierarchy in a top-down structure with simpler subproblems. After the hierarchy is established, alternatives are compared in pairs under one or multiple criteria chosen by decision makers, and the solutions to the subproblems will be aggregated to obtain the final answer to the original problem under study. The AHP hierarchy may consist of many levels in which elements need to be compared pairwise. To avoid the inconsistency introduced by conflictive human judgments, the AHP introduces an eigenvector-based approach to check consistency. If the inconsistency cannot be tolerated, the pairwise comparison step will be performed repeatedly until comparison results are consistent. This feature overcomes the weakness of contradictory human knowledge and significantly improves the subjective comparisons and hence makes the AHP a popular MCDM tool which is widely used in practice.

In the literature, there are numerous applications of the AHP in many different areas such as supply chain management (Ramanathan 2013; Govindan et al. 2014), logistics (Barker and Zabinsky 2011), multisensor data fusion (Frikha and Moalla 2015), manufacturing (Sato et al. 2015), and data analysis (Chan et al. 2015). Moreover, research efforts are conducted to extend the power of the AHP. For example, Dong et al. (2013) proposed a new framework based on the 2-tuple linguistic modeling of AHP scale problems so that decision makers can use to generate numerical scales individually; Durbach et al. (2014) integrated the AHP with stochastic multicriteria acceptability analysis (SMAA) to allow uncertain pairwise comparisons; and Jalao et al. (2014) proposed a beta distribution to model the varying stochastic preferences of decision makers by using the method-of-moments methodology to fit the varying stochastic preferences of the decision makers into beta stochastic pairwise comparisons. All these works make the AHP more powerful and applicable.

Generally, a standard AHP application comprises the five steps below:

1. Hierarchy development: In this step, a top-down hierarchy will be established for the subsequent numerical computation in the AHP. First, it needs to identify the top level with a goal for the problem under study, one or multiple intermediate levels of criteria and subcriteria, and the bottom level which is usually a set of alternatives. Then, the correlated elements in different levels need to be connected to construct a top-down structure.
2. Pairwise comparison: After the hierarchy of an AHP model is established, the elements in each level, except the top goal level, need to be compared pairwise to evaluate their relative significance over others in the same level. Throughout the hierarchy, each element in an upper level will be used to compare the elements in the level immediately below with respect to it. Usually, this step

relies on human knowledge, and a 1–9 scale will be used to measure the relative importance of two elements, which may introduce inconsistency because of contradictory or inaccurate human judgements. A comparison matrix shown in Eq. (12.1) will be obtained from the pairwise comparisons between the elements in each level.

$$W = \begin{bmatrix} \frac{w_1}{w_1} & \frac{w_1}{w_2} & \cdots & \frac{w_1}{w_n} \\ \frac{w_2}{w_1} & \frac{w_2}{w_2} & \cdots & \frac{w_2}{w_n} \\ \vdots & \vdots & \ddots & \vdots \\ \frac{w_n}{w_1} & \frac{w_n}{w_2} & \cdots & \frac{w_n}{w_n} \end{bmatrix}, \quad (12.1)$$

where n is the number of the elements which are compared pairwise.

3. Eigenvalue and eigenvector calculation: An eigenvalue and eigenvector can be calculated from a comparison matrix shown in Eq. (12.1), and the values in the eigenvector indicate the significance of the elements which have been compared. Usually, the eigenvector will be normalized before it is aggregated to the final result.
4. Consistency check: For the accuracy of pairwise comparisons, the consistency of each comparison matrix needs to be checked by using consistency index (CI) or consistency ratio (CR). If a comparison matrix is not consistent, the corresponding pairwise comparisons should be repeated until the matrix is consistent.
5. Priority measurement: Usually, the priorities obtained from the pairwise comparisons at a level will be used to weigh the priorities in the level immediately below and then the weighted values of the elements in lower levels will be added to obtain their overall priority. This process needs to be repeated for every element until the final priorities of the alternatives in the bottom level of the AHP are obtained (Saaty 2008).

12.2.2 Data Aggregation and Disaggregation by the AHP

In a statistical analysis, variables can be classified into two types: qualitative and quantitative. Qualitative variables are non-numerical and usually associated with categorical values, while quantitative variables are numerical and their values are applicable for statistical techniques such as regression (Luo and Brodsky 2010). For example, item and color are two attributes which can be represented by qualitative variables, while the sales of a product is a quantitative variable over time. To gain better historical observations for forecasting, the low sales volumes of individual SKUs can be aggregated over qualitative variables. Although data aggregation is not difficult to implement, it is not easy to distribute the aggregate forecast over SKUs or products. However, the AHP provides a straightforward means to split the

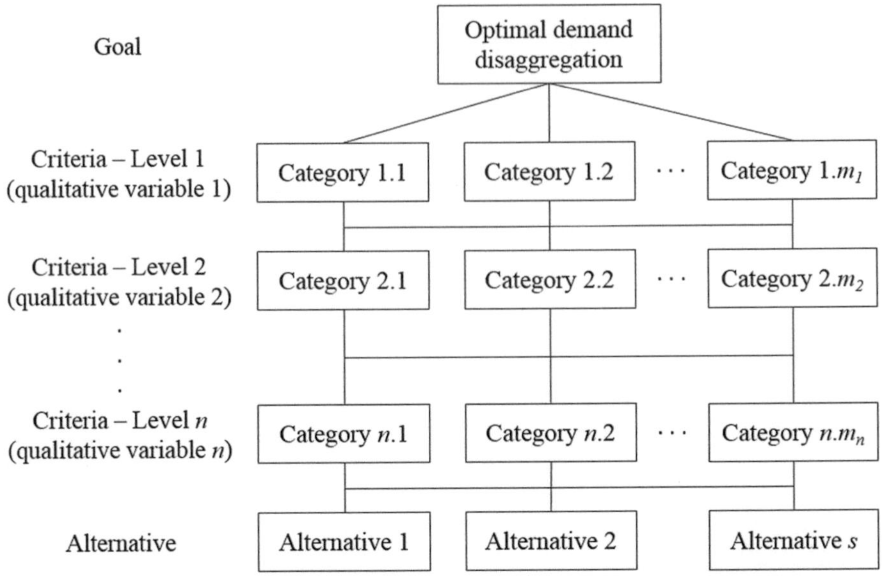

Fig. 12.1 AHP hierarchy for data aggregation and disaggregation in forecasting

aggregate forecast. More specifically, the entities under study are weighted by the AHP based on their historical sales volumes with certainty, and then, the weights are used to distribute the aggregate forecast of future sales over those entities. Generally, the AHP hierarchy for such an application is shown in Fig. 12.1. In practice, the number of the levels of criteria is decided by the number of qualitative variables upon which raw data will be aggregated, and the alternatives can be any qualitative variable of interest such as SKUs or products. In each level, historical sales volumes, either original or aggregate, will be used for pairwise comparisons. Moreover, the element in a level may not be connected to all the elements in the level which is immediate above or below it, and whether a connection exists or not depends on the specific problem under study.

In the fashion industry as well as many other industries, sales data are collected over time and hence attached with time labels, to which time series techniques (Cheng et al. 2015) can be used for sales forecasting. Because of the nature of the sales data, pairwise comparisons in our AHP model will also be conducted by using historical observations at discrete time points. Suppose $t = 1, 2, \ldots, T$ represents discrete time, $n = 1, 2, \ldots, N$ represents criteria levels in the hierarchy, and $m = 1, 2, \ldots, M_n$ $(1 \leq n \leq N)$ represents the elements at criteria level n. Let S be the number of alternatives in the AHP, and $d_s(t)$ $(1 \leq s \leq S,\ 1 \leq t \leq T)$ be the sales of alternative s at time t, then the aggregate sale volume under category m at level n can be calculated by Eq. (12.2),

$$d_m^n(t) = \sum_{s \in \{m_n\}} d_s(t), \quad 1 \le n \le N, \quad 1 \le m \le M_n, \quad 1 \le t \le T, \tag{12.2}$$

where $\{m_n\}$ is the set of alternatives which belongs to category m in level n. Obviously, the bottom level of the AHP, which consists of alternatives, can be regarded as the (N + 1)th level of criteria, so $d_s(t)$ can be reformatted as Eq. (12.3). Thus, for level n, the comparison matrix of the elements at time t can be expressed by Eq. (12.4), and the final comparison matrix can be obtained by Eq. (12.5).

$$d_m^{N+1}(t) = d_m(t), \quad 1 \le m \le S, \quad 1 \le t \le T, \tag{12.3}$$

$$W_n(t) = \begin{bmatrix} 1 & \frac{d_1^n(t)}{d_2^n(t)} & \cdots & \frac{d_1^n(t)}{d_{M_n}^n(t)} \\ \frac{d_2^n(t)}{d_1^n(t)} & 1 & \cdots & \frac{d_2^n(t)}{d_{M_n}^n(t)} \\ \vdots & \vdots & \ddots & \vdots \\ \frac{d_{M_n}^n(t)}{d_1^n(t)} & \frac{d_{M_n}^n(t)}{d_2^n(t)} & \cdots & 1 \end{bmatrix}, \quad 1 \le n \le N+1, \tag{12.4}$$

$$V_n = \sum_{t=1}^{T} a_t W_n(t), \quad 1 \le n \le N+1, \tag{12.5}$$

where a_t is the weight of time period t. If the observations from all time periods are equally weighted, we have $a_t = 1 \ (1 \le t \le T)$. The consistency of $W_n(t)$ is defined as follows.

Definition 1 In Eq. (12.4), $W_n(t)$ is consistent if and only if

$$\frac{d_i^n(t)}{d_j^n(t)} \cdot \frac{d_j^n(t)}{d_k^n(t)} = \frac{d_i^n(t)}{d_k^n(t)}, \quad 1 \le i, j, k \le M_n, \quad i \ne j \ne k.$$

Let $\overrightarrow{u_n} = \left[u_1^n, u_2^n, \ldots, u_{M_n}^n\right]' \ (1 \le n \le N+1)$ be the normalized eigenvector obtained from V_n, then $u_m^n \ (1 \le m \le M_n)$ is the local weight which indicates the significance of category or alternative m at level n comparing to others at the same level. The global weight which estimates the percentage of the sales volumes falling into such category or alternative can be calculated by Eq. (12.6).

$$g_m^n = u_m^n \sum_{m' \subseteq \{1,2,\ldots,M_{n-1}\}} g_{m'}^{n-1}, \quad 1 \le n \le N+1, \tag{12.6}$$

where m' is the set which consists of the parent elements of m at level n – 1.

12.2.3 Adjustment to Exceptional Results of Pairwise Comparison

For any level n and time t, $W_n(t)$ is always consistent because its entries are the results of pairwise comparisons which are conducted on a set of deterministic values. However, the final matrix at level n $(1 \leq n \leq N+1)$, V_n, may not be consistent because the sales quantity of an alternative is a random variable over time (see the proof in Appendix for details). A main power of the AHP comes from its ability to measure and adjust inconsistent pairwise comparisons. In an application where human knowledge is used for pairwise comparisons, the inconsistency can be eliminated by repeating the process conducted by human experts until consistency is achieved. In our scheme, however, inconsistency cannot be adjusted in the traditional way because no human expert will be involved. To make V_n consistent, a possible resolution is to adjust a_t in Eq. (12.5). In case that a_t cannot be adjusted, the entries in V_n $(1 \leq n \leq N+1)$ can also be changed for consistency by using some new values which comprise the comparison results shown in Eq. (12.7).

$$v_{i,j}^n = \frac{\sum_{t=1}^T d_i^n(t)}{\sum_{t=1}^T d_j^n(t)}, \quad v_{j,i}^n = \frac{\sum_{t=1}^T d_j^n(t)}{\sum_{t=1}^T d_i^n(t)}, \quad 1 \leq i, j \leq M_n, \quad i \neq j, \tag{12.7}$$

where $v_{i,j}^n$ is the entry at row i and column j of V_n, and $v_{j,i}^n$ is that at row j and column i.

Another common problem in pairwise comparisons is the occurrence of zero and infinite values. When the sales volume of an alternative or under a category is zero at time t, the related comparison results will be either zero or infinity, and hence, there will be infinite entries in the final matrix accordingly. Since it is not possible to calculate eigenvalues and eigenvectors from such a matrix, the infinite values must be replaced by finite numbers. Mathematically, suppose for element i $(1 \leq i \leq M_n)$ at level n $(1 \leq n \leq N+1)$ during time period t $(1 \leq t \leq T)$, there is $d_i^n(t) = 0$, then there will be $w_{i,j}^n(t) = 0$ and $w_{j,i}^n(t) = \infty$ for any $j \neq i$, where $w_{i,j}^n(t)$ and $w_{j,i}^n(t)$ are the entries in $W_n(t)$. Consequently, there will be $v_{j,i}^n = \infty$, and hence, eigenvalues and eigenvectors cannot be calculated from V_n. In our scheme, we propose the resolution shown in Eq. (12.8).

$$v_{i,j}^n = \frac{1}{b}, \quad v_{j,i}^n = b, \quad 1 \leq n \leq N+1, \quad 1 \leq i, j \leq M_n, \quad i \neq j, \tag{12.8}$$

where b is a constant which can be either predefined or decided by some rules. Intuitively, b should be a large number because it is used to replace the infinity. However, this is not true for the sales forecasting in the fashion industry because of the nature of the business. More specifically, since individual SKUs usually have low sales volumes which may vary significantly, the actual portion of a SKU in the total sales volume can be distorted remarkably if b is very large. For example, if the sales of a high-volume product is zero during a period, which is not unusual in reality, large b values can lead to a small global weight for this product, and consequently, its sales volumes will be under-forecasted. Thus, a small or medium value is suggested for b in this study. In practice, the value of b can be set up by using the knowledge from business experts, or by numerical experiments to find an optimal value which minimizes forecast errors for the specific problem under study.

12.3 Numerical Experimentation

12.3.1 Data Set

In this section, the AHP-based scheme is tested by using real sales data from a fashion boutique in Hong Kong. The data include six fashion items (i.e., T-shirt, dress, bag, pant, accessory, and belt) and seven colors (i.e., black, blue, brown, red, white, green, and gray). Other than item and color, there are three attributes in a sales record: date, quantity, and price. The original data set covers time duration of forty-two weeks in total, and a sample piece is presented in Table 12.1. In this study, the original data are consolidated into a weekly bucket for simplicity, and the sales volumes are aggregated by item to test the scheme we propose. In Table 12.1, since the dates were in two different calendar weeks, the data can be aggregated into two weeks as shown in Table 12.2. There are 42 observations over time in our numerical analysis because the original data set covers 42 calendar weeks. The basic descriptive statistics of those observations under six item categories are provided in Table 12.3.

Table 12.1 Original sales data from a fashion boutique

Date (year-mon-day)	Item code	Color	Quantity	Price
2005-08-06	T-shirt	White	1	10
2005-08-06	Pant	Red	2	50
2005-08-06	T-shirt	Green	1	39
2005-08-07	T-shirt	Black	1	65
2005-08-08	T-shirt	Black	2	79
2005-08-08	Pant	Blue	1	100
2005-08-08	T-shirt	Brown	1	59

Table 12.2 Aggregate sales data by item

Week number	Item	Quantity	Price
1	T-shirt	3	38
1	Pant	2	50
2	T-shirt	3	72.3
2	Pant	1	100

Table 12.3 Global weight of item category (obs = 2, b = 3)

	T-shirt	Dress	Bag	Pant	Accessory	Belt	Sum
Week 37	0.5695	0.0817	0.0417	0.2304	0.0356	0.0411	1.00
Week 38	0.4959	0.1041	0.0405	0.2683	0.0474	0.0438	1.00
Week 39	0.5617	0.0777	0.0396	0.1922	0.0776	0.0513	1.00
Week 40	0.5778	0.0663	0.0534	0.2217	0.0320	0.0488	1.00
Week 41	0.5002	0.0764	0.0509	0.3009	0.0177	0.0539	1.00
Week 42	0.5090	0.0619	0.0608	0.2952	0.0244	0.0487	1.00

Remarks: In this table, obs means the number of historical observations, and b is the parameter specified in Eq. (12.8). This convention will be used in the tables throughout the rest of this paper

12.3.2 Experiment Design and Numerical Analysis

The original data set includes the prices of the products sold in history. It is a quantitative variable upon which many statistical techniques can be applied. However, how to set up this variable in aggregate sales data needs to be considered carefully. Otherwise, the information about the future sales to be predicted may be used implicitly, which will weaken the approach. For example, the average price weighted by sales volumes during a time period should not be used to forecast the sales during the same time period because it contains the information about the sales volumes to be predicted. As an initial research effort, this paper does not consider how to set up quantitative variables when sales data are aggregated. Thus, the AHP-based scheme is only applied to two basic time series forecasting methods, moving average (MA) and exponential smoothing (ES). To test its performance, the sales volumes per item in the last six weeks of the whole time span (i.e., week 37–42) are predicted with and without applying the AHP-based scheme, which means that the forecasting is made by four approaches: MA, MA based on the AHP (MA-AHP), ES, and ES based on the AHP (ES-AHP). In MA-AHP and ES-AHP, the global weights of the six item categories are generated by the AHP model whose hierarchy is shown in Fig. 12.2. Those weights are first generated under several different settings in terms of the number of historical observations and b values, and then are used to distribute the aggregate forecast over individual item categories.

For week t, let F_t be the total aggregate forecast over all SKUs, $d_t^{(i)}$ and $F_t^{(i)}$ be the aggregate sales and forecast volumes under item category i, respectively, and $g_t^{(i)}$ be the global weight of item category i generated by the AHP. Suppose K is the number of the most recent observations used by MA, and α $(0 \leq \alpha \leq 1)$ is the smoothing constant for ES. Then, the four approaches can be expressed by Eqs. (12.9)–(12.12), respectively.

MA:

$$F_t^{(i)} = \frac{1}{K}\sum_{l=t-K}^{t-1} d_l^{(i)}, \quad 1 \leq i \leq 6, \quad 37 \leq t \leq 42. \tag{12.9}$$

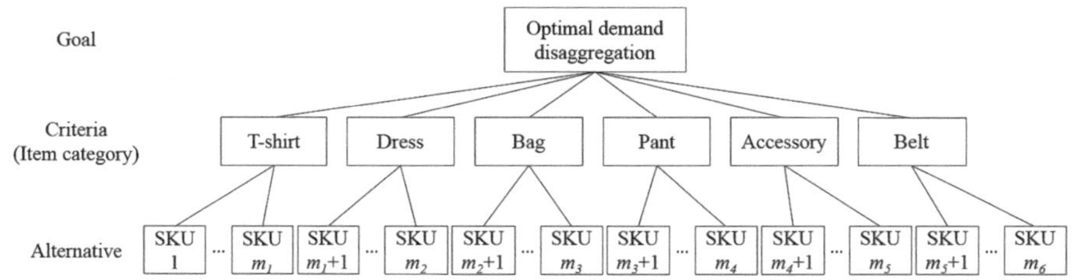

Fig. 12.2 AHP hierarchy for numerical analysis

MA-AHP:

$$F_t = \frac{1}{K}\sum_{l=t-K}^{t-1}\sum_{i=1}^{6} d_l^{(i)}, F_t^{(i)} = g_t^{(i)}F_t, \quad 1 \le i \le 6, \ \ 37 \le t \le 42. \tag{12.10}$$

ES:

$$F_t^{(i)} = \alpha d_{t-1}^{(i)} + (1-\alpha)F_{t-1}^{(i)}, \quad 1 \le i \le 6, \ \ 37 \le t \le 42. \tag{12.11}$$

ES-AHP:

$$F_t = \alpha \sum_{i=1}^{6} d_{t-1}^{(i)} + (1-\alpha)F_{t-1}, \ F_t^{(i)} = g_t^{(i)}F_t, \quad 1 \le i \le 6, \ \ 37 \le t \le 42. \tag{12.12}$$

Tables 12.3, 12.4, and 12.5 illustrate how the AHP-based scheme works by using the most recent two observations for sales forecasting (e.g., observations at weeks 35 and 36 are used for the forecasts at week 37). More specifically, Table 12.3 shows the global weights of the six item categories generated by the AHP, Table 12.4 shows the aggregate forecasts by MA and ES, and Table 12.5 shows the disaggregated forecast quantities per item category.

In the results analysis, the mean squared error (MSE) and the symmetric mean absolute percentage error (SMAPE) are used to measure the sales forecast accuracy of different approaches. MSE is one of the most popular measures which are widely used in both academic research and industrial practices. It is the average of the squared errors which implies how much an estimator deviates from a true value. Mathematically, MSE is defined in Eq. (12.13).

Table 12.4 Aggregate forecast by MA and ES (obs = 2, b = 3)

	Week 37	Week 38	Week 39	Week 40	Week 41	Week 42
MA	51	56	61	67	57	41
ES	51	56	61	67	57	41

Table 12.5 Sales forecast by AHP-MA/AHP-ES (obs = 2, b = 3)

	T-shirt	Dress	Bag	Pant	Accessory	Belt
Week 37	29	4	2	12	2	2
Week 38	28	6	2	15	3	2
Week 39	34	5	2	12	5	3
Week 40	39	4	4	15	2	3
Week 41	29	4	3	17	1	3
Week 42	21	3	2	12	1	2

$$\text{MSE} = \frac{1}{T}\sum_{t=1}^{T}(F_t - A_t)^2, \tag{12.13}$$

where F_t and A_t represent forecast and true value at time t, respectively.

In the fashion industry, it happens from time to time that the sales volume of a product at a retail store is recorded to be zero during a time period. Because of this, there are many zero values in the data set used for our numerical study. Since the mean absolute percentage error (MAPE) is not suitable for handling this scenario, SMAPE, which is a relative measure based on percentage errors, is used as the second measurement of forecast accuracy instead. Mathematically, SMAPE is defined in Eq. (12.14).

$$\text{SMAPE} = \frac{\sum_{t=1}^{T}|F_t - A_t|}{\sum_{t=1}^{T}(F_t + A_t)}, \tag{12.14}$$

where F_t and A_t represent forecast and true value at time t, respectively.

To fully test the performance of the AHP-based scheme, the global weights of the item categories are generated by the AHP model under several different sets of obs and b values. Tables 12.6, 12.7, and 12.8 present the forecast accuracy in terms of MSE and SMAPE under short data history settings. From the numerical results, we can see that the overall performance of both MA and ES can be improved after

Table 12.6 Forecast accuracy of four approaches (obs = 8, b = 4)

		T-shirt	Dress	Bag	Pant	Accessory	Belt
MA	MSE	110	5	2	10	5	4
	SMAPE (%)	15.63	25.58	28.00	9.32	35.71	33.33
MA-AHP	MSE	98	4	2	11	5	2
	SMAPE (%)	14.53	22.73	25.00	10.97	35.71	23.81
ES	MSE	102	4	2	11	6	4
	SMAPE (%)	14.29	25.58	25.00	9.88	40.00	33.33
ES-AHP	MSE	97	4	2	10	5	2
	SMAPE (%)	13.71	22.73	23.08	8.86	35.71	23.81

Table 12.7 Forecast accuracy of four approaches (obs = 8, b = 5)

		T-shirt	Dress	Bag	Pant	Accessory	Belt
MA	MSE	110	5	2	10	5	4
	SMAPE (%)	15.63	25.58	28.00	9.32	35.71	33.33
MA-AHP	MSE	98	4	2	11	5	3
	SMAPE (%)	14.53	22.73	23.08	10.97	35.71	30.00
ES	MSE	102	4	2	11	6	4
	SMAPE (%)	14.29	25.58	25.00	9.88	40.00	33.33
ES-AHP	MSE	97	4	2	10	5	3
	SMAPE (%)	13.71	22.73	25.93	9.43	35.71	30.00

Table 12.8 Forecast accuracy of four approaches (obs = 10, b = 4)

		T-shirt	Dress	Bag	Pant	Accessory	Belt
MA	MSE	109	4	2	11	6	3
	SMAPE (%)	16.17	21.95	30.77	11.25	37.93	30.00
MA-AHP	MSE	105	3	2	13	5	2
	SMAPE (%)	15.29	19.05	28.00	12.42	35.71	23.81
ES	MSE	100	4	2	12	6	4
	SMAPE (%)	14.71	25.58	28.00	10.43	40.00	33.33
ES-AHP	MSE	106	3	2	13	5	2
	SMAPE (%)	14.70	19.05	21.43	11.54	35.71	27.27

the AHP-based scheme is applied. In particular, for item "belt" which has the smallest sales quantity among the six items, the forecast accuracy has been significantly increased in terms of MSE and SMAPE, which indicates that our scheme is an effective means to improve the forecast quality for the fashion products with short life cycle and low sales volumes. From Tables 12.6, 12.7 and 12.8, we can also see that the numerical results from the three experiments are similar, which indicates that the AHP-based scheme works consistently under similar parameter settings. Moreover, for item "belt," there is a significant change in forecast accuracy when b changes, which indicates that low-volume items are more sensitive to b comparing to high-volume items.

12.4 Conclusion and Future Work

Sales forecasting is a very challenging problem in the fashion industry. Although the wise use of information for conducting sales forecasting will be greatly helpful to enhance the operations management of fashion companies (Mishra et al. 2009), it is not easy to do so because the fashion products exhibit the features of short life cycles, low sales volumes, and significant volatility. In this paper, we propose a novel scheme based on the AHP to aggregate sales data for "better historical observations" on which the total future sales of multiple products (or SKUs) can be predicted more accurately. The aggregate forecast can then be distributed over the business unit of interest to capture the future demand of those units more effectively. Notice that in the literature, many research efforts are conducted to improve a certain type of techniques for forecasting. For example, a multistep expectation maximization based algorithm was proposed by Luo et al. (2012) to improve piecewise surface regression for a better forecast quality. But unlike those works, this paper proposes a general framework which is developed to make a better use of historical data for forecasting instead of improving any specific techniques afterward for sales forecasting. Thus, it can be applied to help most existing approaches to enhance their performance in sales forecasting.

Future studies can be conducted in the following directions. First, as mentioned in Sect. 12.3.2, how to set up quantitative variables when sales data are aggregated has not been studied in the literature. By solving this problem, it will be possible to apply the AHP-based schema to many advanced forecasting techniques and hence the power of the method will be well extended. Second, the constant b in Eq. (12.8) is a predefined parameter in the numerical experiments presented in this paper; it is observed that the forecast accuracy may be significantly affected by the value of this parameter. To optimize the performance of the scheme, it will be greatly beneficial to develop a method which decides the optimal value of this parameter b. Last, this scheme is of great value in practice because of its runtime efficiency. Theoretically, this scheme should reduce the runtime of statistical approaches significantly because the number of statistical analysis will be decreased a lot when a large number of SKUs are aggregated together (as the runtime of the AHP model is very short comparing with statistical models). Thus, in an industrial application where data volume is extremely high, our proposed scheme will be highly preferred because of its great capability in runtime reduction even if it cannot improve forecast quality. However, the time performance of the AHP-based scheme cannot be tested in this study because we need much more data to do so. We thus relegate this extension to our future research.

Acknowledgement We are sincerely grateful to Dr. Tsan-Ming Choi, Professor of Fashion Business at The Hong Kong Polytechnic University, for providing the sales data for our numerical study. Also, we would like to thank the editor and referees for their kind advice and comments.

Appendix

Lemma *The final comparison matrix in* Eq. (12.5) *can be inconsistent if it is aggregated from the comparison matrices in* Eq. (12.4).

Proof Suppose that at criteria level n, i, j, and k are three elements which are compared pairwise. Without loss of generality, let us assume that pairwise comparisons are conducted on the observations at two discrete times, namely 1 and 2, for which $a_1 = a_2 = 1$ is satisfied. Thus, Eqs. (12.15)–(12.17) can be obtained.

$$v_{i,j}^n = \frac{d_i^n(1)}{d_j^n(1)} + \frac{d_i^n(2)}{d_j^n(2)} = \frac{d_i^n(1)d_j^n(2) + d_i^n(2)d_j^n(1)}{d_j^n(1)d_j^n(2)}, \tag{12.15}$$

$$v_{i,k}^n = \frac{d_i^n(1)}{d_k^n(1)} + \frac{d_i^n(2)}{d_k^n(2)} = \frac{d_i^n(1)d_k^n(2) + d_i^n(2)d_k^n(1)}{d_k^n(1)d_k^n(2)}, \tag{12.16}$$

$$v_{j,k}^n = \frac{d_j^n(1)}{d_k^n(1)} + \frac{d_j^n(2)}{d_k^n(2)} = \frac{d_j^n(1)d_k^n(2) + d_j^n(2)d_k^n(1)}{d_k^n(1)d_k^n(2)}. \tag{12.17}$$

If the final comparison matrix in Eq. (12.5) is consistent, then there we have the following: $v_{i,j}^n \cdot v_{j,k}^n = v_{i,k}^n$. In this case, Eq. (12.18) must be true. Since sales volumes are random and Eq. (12.18) cannot be always satisfied, the lemma is proved.

$$d_i^n(1)d_j^n(2)d_j^n(2)d_k^n(1) + d_i^n(2)d_j^n(1)d_j^n(1)d_k^n(2) = 0. \tag{12.18}$$

References

Albey E, Norouzi A, Kempf KG, Uzsoy R (2015) Demand modeling with forecast evolution: an application to production planning. IEEE Trans Semicond Manuf 28(3):374–384

Asian S, Nie X (2014) Coordination in supply chains with uncertain demand and disruption risks: existence, analysis, and insights. IEEE Trans Syst Man Cybern Syst 44(9):1139–1154

Au KF, Choi TM, Yu Y (2008) Fashion retail forecasting by evolutionary neural networks. Int J Prod Econ 114:615–630

Barker TJ, Zabinsky ZB (2011) A multicriteria decision making model for reverse logistics using analytical hierarchy process. OMEGA Int J Manage Sci 39:558–573

Bruzzone A, Longo F, Nicoletti L, Chiurco A, Bartolucci C (2013) Multiple forecasting algorithms for demand forecasting in the fashion industry. In: Proceedings of the 8th EUROSIM congress on modelling and simulation (EUROSIM), pp 421–426

Chan HK, Wang X, Lacka E, Zhang M (2015) A mixed-method approach to extracting the value of social media data. Prod Oper Manage (published online)

Chen T, Wang YC (2014) An agent-based fuzzy collaborative intelligence approach for precise and accurate semiconductor yield forecasting. IEEE Trans Fuzzy Syst 22(1):201–211

Cheng C, Sa-Ngasoongsong A, Beyca O, Le T, Yang H, Kong Z, Bukkapatnam S (2015) Time series forecasting for nonlinear and non-stationary processes: a review and comparative study. IIE Trans 47:1053–1071

Choi TM, Hui CL, Ng SF, Yu Y (2012) Color trend forecasting of fashionable products with very few historical data. IEEE Trans Syst Man Cybern Appl Rev 42(6):1003–1010

Choi TM, Hui CL, Liu N, Ng SF, Yu Y (2014) Fast fashion sales forecasting with limited data and time. Decis Support Syst 59:84–92

Cui S, Zhao M, Ravichandran T (2011) Market uncertainty and dynamic new product launch strategies: a system dynamics model. IEEE Trans Eng Manage 58(3):530–550

Cui RM, Allon G, Bassamboo A, Van Mieghem JA (2015) Information sharing in supply chains: an empirical and theoretical valuation. Manage Sci 61(11):2803–2824

Danese P, Kalchschmidt M (2011) The role of the forecasting process in improving forecast accuracy and operational performance. Int J Prod Econ 131:204–214
Dong Y, Hong WC, Xu Y, Yu S (2013) Numerical scales generated individually for analytic hierarchy process. Eur J Oper Res 229:654–662
Du W, Leung YS, Kwong CK (2015) A multiobjective optimization-based neural network model for short-term replenishment forecasting in fashion industry. Neurocomputing 151:342–353
Durbach I, Lahdelma R, Salminen P (2014) The analytic hierarchy process with stochastic judgements. Eur J Oper Res 238:552–559
Frikha A, Moalla H (2015) Analytic hierarchy process for multi-sensor data fusion based on belief function theory. Eur J Oper Res 241:133–147
Govindan K, Kaliyan M, Kannan D, Haq AN (2014) Barriers analysis for green supply chain management implementation in Indian industries using analytic hierarchy process. Int J Prod Econ 147:555–568
Heinecke G, Syntetos AA, Wang W (2013) Forecasting-based SKU classification. Int J Prod Econ 143:455–462
Hu CH, Si XS, Yang JB, Zhou ZJ (2011) Online updating with a probability-based prediction model using expectation maximization algorithm for reliability forecasting. IEEE Trans Syst Man Cybern Syst Hum 41(6):1268–1277
Jalao ER, Wua T, Shunk D (2014) A stochastic AHP decision making methodology for imprecise preferences. Inf Sci 270:192–203
Jin Y, Williams BD, Tokar T, Waller MA (2015) Forecasting with temporally aggregated demand signals in a retail supply chain. J Bus Logistics 36(2):199–211
Kaya M, Yesil E, Dodurka MF, Siradag S (2014) Fuzzy forecast combining for apparel demand forecasting. In: Choi TM, Hui CL, Yu Y (eds) Intelligent fashion forecasting systems: models and applications. Springer, Berlin, pp 123–146
Liu N, Ren S, Choi TM, Hui CL, Ng SF (2013) Sales forecasting for fashion retailing service industry: a review. Math Prob Eng 2013. Article ID 738675
Luo J, Brodsky A (2010) An optimal regression algorithm for piecewise functions expressed as object-oriented programs. In: Proceedings of the 9th International conference on machine learning and applications (ICMLA), 937–942
Luo J, Brodsky A, Li Y (2012) An EM-based ensemble learning algorithm on piecewise surface regression problem. Int J Appl Math Stat 28(4):59–74
Luo M, Yan HC, Hu B, Zhou JH, Pang CK (2015) A data-driven two-stage maintenance framework for degradation prediction in semiconductor manufacturing industries. Comput Ind Eng 85:414–422
Mishra BK, Raghunathan S, Yue X (2009) Demand forecast sharing in supply chains. Prod Oper Manage 18(2):152–166
Mostard J, Teunter R, Koster R (2011) Forecasting demand for single-period products: a case study in the apparel industry. Eur J Oper Res 211:139–147
Nenni ME, Giustiniano L, Pirolo L (2013) Demand forecasting in the fashion industry: a review. Int J Eng Bus Manage 5:1–6
Ramanathan U (2013) Aligning supply chain collaboration using analytic hierarchy process. OMEGA Int J Manage Sci 41:431–440
Ren S, Choi TM, Liu N (2015) Fashion sales forecasting with a panel data-based particle-filter model. IEEE Trans Syst Man Cybern Syst 45(3):411–421
Saaty TL (1980) The analytic hierarchy process: planning, priority setting, resource allocation. McGraw-Hill, New York
Saaty TL (2008) Decision making with the analytic hierarchy process. Int J Serv Sci 1(1)
Sato Y, Tan KH, Tse YK (2015) An integrated marginal analysis approach for build-to-order products. Int J Prod Econ 170:422–428
Schwartz JD, Arahal MR, Rivera DE, Smith KD (2009) Control-relevant demand forecasting for tactical decision-making in semiconductor manufacturing supply chain management. IEEE Trans Semicond Manuf 22(1):154–163
Sen A (2008) The US fashion industry: a supply chain review. Int J Prod Econ 114(2):571–593

Subramanian N, Ramanathan R (2012) A review of applications of Analytic Hierarchy Process in operations management. Int J Prod Econ 138:215–241

Sun ZL, Choi TM, Au KF, Yu Y (2008) Sales forecasting using extreme learning machine with applications in fashion retailing. Decis Support Syst 46:411–419

Thomassey S (2010) Sales forecasts in clothing Industry: the key success factor of the supply chain management. Int J Prod Econ 128:470–483

Wong WK, Guo ZX (2010) A hybrid intelligent model for medium-term sales forecasting in fashion retail supply chains using extreme learning machine and harmony search algorithm. Int J Prod Econ 128:614–624

Yu Y, Choi TM, Hui CL (2011) An intelligent fast sales forecasting model for fashion products. Expert Syst Appl 38:7373–7379

Yu Y, Hui CL, Choi TM (2012) An empirical study of intelligent expert systems on forecasting of fashion color trend. Expert Syst 7Appl 39:4383–4389

Index

T.-M. Choi (ed.), *Analytical Modeling Research in Fashion Business*,
Springer Series in Fashion Business, DOI 10.1007/978-981-10-1014-9

MIX
Papier aus verantwortungsvollen Quellen
Paper from responsible sources
FSC® C105338

If you have any concerns about our products,
you can contact us on
ProductSafety@springernature.com

In case Publisher is established outside the EU,
the EU authorized representative is:
Springer Nature Customer Service Center GmbH
Europaplatz 3, 69115 Heidelberg, Germany

Printed by Libri Plureos GmbH
in Hamburg, Germany